"There is no suggestion in *Anthropause* that managed degrowth will be easy, predictable, or uncontested—or that any one person will have the blueprint. But revolutionary change doesn't happen without a shared vision, and Stan Cox moves us closer by supplying one that is doable and humane. This book is welcome inspiration for those of us tired of climate doom-stories and delusional growth fiction. We could all do with an *Anthropause*."

—ELEANOR BOYLE, author of *Mobilize Food!: Wartime Inspiration for Environmental Victory Today*

"For anyone who wonders how changing our economic system could radically improve our lives— read this book! *Anthropause* combines an almost utopian vision with concrete examples and policies based firmly in scientific research that could make life better for all of us. Stan Cox offers an accessible and inspiring vision."

—AVIVA CHOMSKY, author of *Is Science Enough?: Forty Critical Questions about Climate Justice*

Anthropause

The Beauty of Degrowth

Stan Cox

SEVEN STORIES PRESS

NEW YORK • OAKLAND • LONDON

Seven Stories Press
140 Watts Street
New York, NY 10013
www.sevenstories.com

Library of Congress Cataloging-in-Publication Data is on file.

ISBN: 978-1-64421-514-2 (hardcover)
ISBN: 978-1-64421-515-9 (ebook)

College professors, high school, and middle school teachers may order free examination copies of Seven Stories Press titles. Visit https://www.sevenstories.com/pg/resources-academics or email academic@sevenstories.com.

Printed in the USA.

9 8 7 6 5 4 3 2 1

Contents

The Days the Earth Stood Still

In spring 2020, a large share of the world's people found themselves confined at home, either voluntarily or in response to lockdown orders. The goal, of course, was to slow transmission of the coronavirus causing Covid-19, but there were other consequences, and some of them turned out to be pleasant surprises.

After more than a century of exponential increase, global carbon emissions plunged 8.8 percent worldwide. Reductions were greatest in some of the highest-emitting regions: 13 percent in the United States and Europe, 12 percent in Brazil, and more than 15 percent in India. Walking and cycling increased dramatically. Air and noise pollution levels plummeted in large cities. Thanks to reductions in industrial activity and tourism, streams, rivers, and lakes became cleaner.[1]

Even more striking was how animal life quickly filled spaces that humans had suddenly deserted. Enchanting photos flooded in from around the world. A herd of wild sheep and goats grazed alongside the expressway leading into Istanbul's international airport. Mountain goats window-shopped on a main street in Wales; a deer strode through a striped crosswalk, Abbey Road–style, in Nara, Japan. Fallow deer grazed on lawns in East London. A herd of water buffalo took command of a major highway in New

Delhi. A sea lion lounged on a Buenos Aires sidewalk, while a puma prowled through downtown Santiago, Chile. Raccoons romped in the surf at a public beach in Panama.[2] Scientists who studied and documented this flourishing of nature in real time began referring to the phenomenon as an "anthropause."[3]

As animals roamed freely, the pandemic took a devastating toll on human life. Further hardship resulted from the very measures that were necessary to limit the number of people infected or killed. Alongside the misery, however, came some positive changes in people's lives. A survey of more than three thousand adults under lockdown in Scotland found that many were enjoying more quality time and improved relationships with partners, family, and friends. Most had been exercising more than usual and paying better attention to their health. A whopping 83 percent had become "more appreciative of things usually taken for granted."[4]

In August 2020, with the second surge of the coronavirus underway, the Pew Research Center asked more than nine thousand US adults about their pandemic experience so far. As one would expect, a solid majority of the comments were negative, with 89 percent of those surveyed mentioning at least one unwelcome change in their lives. But 73 percent mentioned at least one unexpected positive change. Pew published verbatim quotes from some of the people they interviewed. The variety of responses is striking; furthermore, the lion's share of people's quotes, both negative and positive, referred not to the experience of the disease itself but to the measures they took to protect themselves from it.[5]

To me, the positive side effects of efforts to suppress transmission of the coronavirus clearly suggest that if humanity adopts radical measures to reduce greenhouse gas emissions to near zero on an expedited schedule, many other problems might be ameliorated. To be sure, Covid-19 and climate change are very different kinds of problems. The former struck suddenly and hard, while the latter has unfolded slowly over decades, with dramatic impacts becoming clearly apparent only in recent years. Nevertheless, because affluent countries have procrastinated for three-plus decades and carbon has continued accumulating in Earth's atmosphere, the need for forceful climate action is as immediate and urgent as was the need to suppress coronavirus transmission in spring 2020. The measures required to rein in greenhouse gas emissions are of much greater depth and breadth than those necessitated by the pandemic and will have an even broader range of consequences. But I anticipate that among those repercussions, we'll find plenty of beautiful surprises.

THE GOOD, THE BAD, AND THE UPSHOT

Over the past two centuries, industrial and postindustrial enterprises have been yanking societies back and forth between an improved and a degraded quality of life. Industry has developed preventatives and cures that increase lifespans but have also filled our world with toxic pollution, overmedication, and lethal weaponry, all of which shorten people's lives. Industry has produced enough foodstuff per capita to keep everyone well fed—in theory—but billions suffer routine hunger while billions of others are oversupplied with unhealthful food. Corporations have made it possible—for privileged people at least—to move from place to place faster than any other organism on earth, but at a catastrophic cost to the environment and social justice. Big corporations have made intercontinental travel possible for hundreds of millions while providing the aircraft, ships, armored vehicles, missiles, and bombs that keep armed conflict raging somewhere on the globe every day of the year. They've developed systems that give us access to information from across the world in billionsfold greater quantities than before. Still, those systems have proven far more efficient in spreading disinformation, propaganda, and hate than in bringing us constructively together. All this economic production—the harmful, every bit as much as the useful—is aimed at keeping alive the impossible dream of exponential economic growth without end.

If we move boldly to prevent ecological collapse by deeply altering our values and curbing our material production and consumption, we will reap a host of other societal, ecological, and personal benefits that will enrich us with a more beautiful and meaningful life. In the process, I'll show, we can free ourselves from a multitude of harms that plague us in today's growth-at-any-cost society.

The public health policies of 2020 certainly had some beneficial side effects. Anthropologists, geographers, and other academics seemed especially inspired by an apparent resurgence of interest in humanity's relationship with nature. A decades-long trend in which Americans had been spending more and more time indoors was reversed almost overnight.[6] People took to gardening, hiking, bird-watching, cooking out, and playing backyard games. Whereas many national and state parks closed their gates, many urban parks and other communal green spaces stayed open, and they were in high demand.[7]

In a 2020 study, 83 percent of US and UK survey subjects reported feeling an enhanced love of life and living things, an emotion that the surveyors, citing Edward O. Wilson, labeled *biophilia*, "the innate emotional affiliation that humans have with non-human life forms."[8] Even those respondents with the lowest interest in nature before the pandemic experienced a greater sense of biophilia during it.[9]

Noting that the Global North's retreat from "relentless consumerism and 'always on' economies" during the pandemic had immediate ecological benefits, a group of scholars wondered if the deadly pandemic might prompt the world's privileged to pull back from the long-term damage we are doing to the earth and be better off for it. Urging that we take from the anthropause a lesson that a "fundamental scaling back of high-impact activities" can accomplish the kind of ecological healing that technology cannot, they argue that the "anthropause metaphor also implies that there are physical and spiritual benefits to slowing down, treading lightly, and limiting unnecessary human interventions in natural processes. We note that these associations are consistent with non-Western and Indigenous conceptualizations of human-nature relationships as requiring care, balance and reciprocity."[10]

Other academic observers hoped we would be prompted to "replace a sense of owning with a sense of belonging."[11]

DEGROW OR DESTROY

In 1973, the ecological economist Herman Daly published his book *Toward a Steady-State Economy*, making the case for an upper limit on material-resource use and economic activity. The idea was accepted and elaborated upon by other heretical economists and environmentalists in the decades that followed. However, as we approached the tipping point toward irreversible ecological catastrophe, it became increasingly clear to a broader range of researchers and analysts that economic growth must not only be halted but also reversed.

A year before Daly's book was published, the Austrian philosopher André Gorz coined the term *décroissance*, referring to the degrowth of material production.[12] Interest in degrowth gradually increased over the decades that followed, but starting in the early 2010s, the degrowth movement began to spread widely. Degrowth is now the subject of hundreds of articles in academic journals, shelves full of books, and the open-source *Degrowth Journal*.[13]

The degrowth scholar Giorgos Kallis, of the Universitat Autònoma de Barcelona, has defined *degrowth* as "a trajectory where the 'throughput' (energy, materials and waste flows) of an economy decreases while welfare, or well-being, improves."[14] In his 2018 book *Degrowth*, he listed nine principles that he foresaw underlying a degrowth society. Thus, as he and others envisioned such a society, land and labor are treated as being outside the economy. In this vision, the economy is nonexploitative and radically egalitarian. Infrastructure, resources, and goods and services related to health, education, water, and energy are treated as a commons.[15] Production is localized. The economy is diverse, with cooperatives and non-profits predominating and production for markets confined to the far smaller role that it had in precapitalist civilizations. Resources are withdrawn from wasteful and superfluous material production, with some redirected as needed toward relational goods such as "friendship and love, healthy relationships, kinship, etc." Governance occurs through direct democracy to the extent possible, with the beneficial side effect that "spending time and resources on [direct] democratic politics rather than productive investments will in turn slow down the economy." Economic surplus is used for expenditures that are collective, egalitarian, and often recreational, with low energy and material input. Care work, more equally shared between genders, is prioritized; crucially, caring should reach beyond fellow humans to "other living beings and species."[16]

Degrowth stands in stark opposition to the "green growth" argument that physical and economic expansion is not subject to biophysical limits and can be rendered ecologically benign through technological means. And the degrowth vision is prevailing. A 2023 survey of 789 climate-policy researchers found almost three-quarters of them favoring degrowth or no-growth over green growth.[17] Voluminous scholarly literature debunking green growth continues to apply despite its recent rebranding as "the abundance agenda."[18]

Because the degrowth vision is radical in the best sense of the word, the governments of today's Global North are militantly opposed to it. It's nowhere on their agendas, which means that it's even more critical for those who want a livable future to advocate for the degrowth vision. The term *degrowth* has been criticized for what is interpreted as its negative vibe and lack of appeal in cultures that value the many positive connotations of the word *growth*. Addressing that criticism in his 2019 doctoral thesis, "The Political Economy of Degrowth," Timothée Parrique explained that degrowth is intentionally provocative:

> If sustainable development or green economy advertise a certain vision of prosperity, degrowth *subvertises* it; it hijacks the notion of growth. The word itself creates dissensus [the opposite of consensus]; it acts as a semantic weapon of mass disruption, as conceptual dynamite shaking the foundation of growthism and making space for discussion.[19]

Having tracked down at least fifty-eight definitions of *degrowth*, Parrique grouped them into three types. The first group—"environmentalist"—describes "degrowth as decline." This type puts "emphasis on that which should be reduced, e.g., production, material and energy consumption, economic activities, throughput, or anything else linked to environmental pressure."[20] An example of this type of definition is "the intentional limiting and downscaling of the economy to make it consistent with biophysical boundaries."[21] This is the definition of *degrowth* that its critics typically have in mind, and it's not an easy idea to sell on its own. These days, it's almost always discussed in conjunction with the other two faces of degrowth.

What Parrique calls the second, "revolutionary" type of definition, or "degrowth as emancipation," stresses the societal ills, damages, and injustices that result from material overproduction and overconsumption and the relief that degrowth will provide.[22]

The third type is "the utopian definition," "degrowth as destination," or as Parrique also puts it, "degrowth *for* or *to* something . . . In this aspirational understanding of the term, degrowth is associated to a variety of desirable values (e.g. well-being, frugality, justice, sustainability, conviviality, freedom, democracy) that are to be achieved via a decline, an emancipation, or both." Among the "degrowth as destination" definitions on his list is this: "a democratically led, proportional and redistributive downscaling of production and consumption as a means to achieve environmental sustainability, social justice and wellbeing."[23]

The chief focus of this book is on his second aspect, degrowth as emancipation. That is, I examine the many ways we can free ourselves from the growth economy and its many harmful impacts by seeing true wealth as the collective pursuit of meaning, social justice, and beauty while living within ecological limits. I should, however, point out Parrique's observation that "the difference between *emancipation* and *destination* is a thin one" because "the negation of something always comes with the affirmation of something else."[24] Accordingly, in the coming chapters, I'll show how our emancipation from many of the harms currently inflicted by unrestrained growth will open up space for greater well-being in other, sometimes unexpected parts of our lives.

Leading degrowth advocates are explicit in urging that this vision of the future should not be built around hierarchical decision-making. The transformation must be a collective, lowercase *d* democratic effort, rooted in sociopolitical transformation rather than technological solutionism, with mutual aid a constant. Over time, degrowth has also become more explicitly anti-capitalist. Céline Keller, in the graphic nonfiction work *Who Is Afraid of Degrowth?*, writes, "Degrowth is a temporary phase to dismantle capitalism and reverse uneconomic growth, a transition at the end of which stands a well-being, postgrowth economy that is smaller and slower but fits within its biocapacity."[25] With less industrial production will come the need and desire for highly equitable distribution of economic and political power.

Degrowth is also anticolonial and antiracist; proponents stress that only the world's economically dominant economies need to shrink their material production and that people in low-income, dominated countries *really, really* need for the dominant countries to degrow if the South is to have a livable future. Kallis puts it this way: "Economic growth is a colonial idea. It's the way the West found to export its ideology and indirectly control the rest of the world. Degrowth wants an end to all that crap."[26]

Most of the research, discussion, and writing about degrowth so far has come out of academia and radical politics. Putting degrowth into practice in the North will require that it be adopted as the goal in a region or nation somewhere, and that can happen only after political pressure and biophysical realities converge to render the continued pursuit of growth impossible. At that point, it will be imperative to achieve degrowth through a rational, humane process. Although we cannot predict in any detail what day-to-day life in a society that functions under Kallis's nine principles of degrowth will be like, we do know what life in, say, the present-day United States is like, and we all have our ideas about what is bad and getting worse as we follow the business as usual trajectory. That brings us back to the thought experiment at the center of this book: If the United States were to abandon that trajectory and take the concrete actions that are needed to achieve equitable ecological renewal, what elements of life in the present-day Global North would we necessarily, and gladly, leave behind?

GETTING ACROSS THE STARTING LINE

Degrowth advocates rarely elaborate on specific political processes by which degrowth can be achieved or what the new society will look like in its details. The experts in the field acknowledge that its sweeping vision is very far from today's reality, so the struggle for degrowth is sure to play out in highly unpredictable ways.

Given all that, many people respond to ideas of degrowth with fear of the unknown, as they cling to growth as the "devil they know." My aim is to counteract whatever nostalgia we might feel for the growth economy by emphasizing the many facets of our lives that will be enhanced.

Imagining ourselves in a degrowth future, of course, begs the question of how that better world came to be. The best answer I can come up with is that a concerted, thoroughgoing effort to drive greenhouse gas emissions down to zero, reduce material resource use, and prevent mass extinction (absolutely essential actions encompassed by Parrique's "environmental" definition of *degrowth*) will require that we as a society adapt to a smaller energy and material resource supply in ways that ensure material sufficiency and equity (but not excess) for all. Those adaptations, in turn, can provide the footing for Parrique's two other visions of degrowth: "as emancipation" and "as destination."

To imagine a mechanism for jump-starting progress toward degrowth, let's turn to a proposal for *directly* suppressing greenhouse gas emissions: the "cap and adapt" framework that Larry Edwards and I have outlined for phasing out oil, gas, and coal in the United States.[27] One could pursue other approaches, but when envisioning degrowth as emancipation and destination, I will use this one policy proposal as an example of how to embark on the transition. The most directly targeted policy is a fossil-fuel phaseout.

Among the many environmental crises spun off by economic growth, the headliners are human-induced heating of the earth and an accelerating loss of biodiversity. Because climate change is a major cause of biodiversity loss (though far from the only one), cutting greenhouse gas emissions is an especially high priority. And because three-fourths of human-caused emissions are produced by burning fossil fuels, the industries that extract the fuels should be the top targets for climate action. In the United States, as in other rich nations, corporations hold the economic and political systems hostage, so the federal government has mostly left oil, gas, and coal alone—except when they're actively supporting them. They've focused instead on market-based "solutions" featuring alternative-energy technologies, and those approaches are failing to reduce emissions at significant rates.[28]

Fossil capitalism must be confronted head-on with a rapid phaseout of oil, gas, and coal emissions. If we don't do that, we're cooked. Emissions from land-use practices, aviation, and other sources also must be addressed, and reversing all serious threats to the ecosphere will require many actions beyond eliminating greenhouse gas emissions. However, because a fuel phaseout will reduce the total-energy supply, it will help reduce many kinds of ecological damage that go far beyond global warming. The exit of fossil fuels will also create a pressing need for additional policies that foster humane, egalitarian degrowth.

Let's imagine, then, a future United States in which national caps are placed on the numbers of barrels of oil, cubic feet of gas, and tons of coal that may be extracted and burned annually and that those caps are ratcheted down rapidly year by year. That will trigger seismic adjustments throughout society. For one thing, a deep transformation of our built environment and transportation systems is required. And because the fuel phaseout would be so rapid that the buildup of renewable-energy capacity won't be able to keep pace, the economy will have to operate on less total energy per capita. Adaptation will require that energy and material resources be allocated toward meeting society's basic needs and that fuels and electricity be rationed. Those measures, in turn, will trigger socioeconomic adaptations, chief among them guaranteed access to universal services. Such services will include, among other elements, publicly run water and energy utilities; health services; public education and transportation; a nutritious food supply; good-quality housing; ample green space; freedom from pollution; and public safety without repression. [29]

The phaseout of oil, gas, and coal in affluent and high-emitting countries will reduce industrial capacity and slow other economic activities, leading to less extraction and use of resources, less material production, and more ecological recovery like we saw during the Covid-19 anthropause of 2020. The suppression of fossil fuel, through curtailment of industrial production, would create a necessity—what I see as an opportunity—to stop producing goods and services that do more harm than good. That will reverse an ecologically unsustainable trend that has been moving in parallel and often connecting with the climatic and biodiversity emergencies: growth in the industrial exploitation of biophysical resources.

Civilization's material footprint, from extraction, production, construction, and pollution, cannot sustain its current trajectory of expansion. The numbers are shocking. For one thing, the size of that footprint is doubling every twenty years. Consequently, sometime around 2021, the global quantity of human-made mass—that is, the total weight of inanimate solid objects manufactured or constructed by humanity and still intact—surpassed the total weight of all living plant, animal, and microbial biomass on Earth.[30]

It's no secret where the reduction in humanity's material resource consumption needs to happen. The UN Environment Programme maintains a Global Material Flows Database covering almost two hundred nations. The twenty lowest-income nations on the list have a material footprint of 3.4 metric tons per person per year. The footprint of the twenty nations ranking in the middle for income is 9.7 tons per person, and for the richest twenty nations, it's 36.5 tons per person, more than ten times as large as the poorest nations' footprints.[31]

A fossil-fuel phaseout will restrict rich economies' ability to produce or import such huge masses of material goods (or generate proportionately enormous quantities of waste). But more efforts to directly reduce material flows will be needed to grapple with and subdue industrial overproduction. First, manufacturing and construction will need to be aimed at meeting basic human needs, not generating profits. Once universal, equitable access to food, water, shelter, and health care is established, remaining resources will be used for the collective good but not for wasteful or superfluous production.

These goals can be achieved through policies such as the following:

o   Reallocate energy and material resources toward satisfaction of basic needs.

o   Reject nonessential technologies and infrastructure that depend on high-energy consumption or other resource use.

o   Reconfigure residential, commercial, and transportation patterns to end disparities, improve quality of life for all, and deeply reduce reliance on private vehicles.

o   Achieve much greater equality of travel, with very deep reductions in travel by the affluent.

o   Overhaul food production, processing, and distribution, taking agriculture back to its roots.

o   Free us all from employment that is oppressive and/or pointless, wastes resources, causes ecological harm, or benefits only the employer.

o   Strive for a twenty-hour workweek in which conviviality, not profit lust and exploitation, is the rule.

Under these policies and others that ensure sufficiency for all, the sharing of resources can be organized through various forms of collective decision-making, community by community. Each will be working within similar overall boundaries, but how those fair shares are implemented within those boundaries would be determined locally. There can also be means of establishing pools of collective fuel and electricity credits to be allocated by the local community for the common good. A radical transformation of labor relations and the workplace, along with universal basic income and services, will be essential. Care work, more equally shared between genders, will be prioritized; crucially, caring should reach beyond fellow humans to other life on Earth. All these policies are prominent in the degrowth vision.

The above scenario will also entail a transformation of how we produce food and other biological necessities (see chapter IV). It will take aim at the 25 percent or so of greenhouse gas emissions that are created by agricultural production and processing, forest exploitation, and other uses of the earth's lands and waters. And that's not all: current land use is also a major, direct threat to biodiversity.

Obviously, we can't phase out food production. Fortunately, almost all the changes required to reduce greenhouse gas emissions from farming and forest use are already badly needed for a wide range of other ecological and human-welfare reasons. They would be crucially important steps to take even if global warming were not a problem.[32]

It probably goes without saying—but nevertheless should be said—that the possibility of achieving all the above, from the fossil-fuel phaseout to the planned allocation of resources to the overhaul of agriculture, seems further out of reach than ever in mid-2020s America. Corporations, financial institutions, the military-industrial complex, and monomaniacal billionaires now wield greater power over our society than ever before. They will instantly veto any effort to phase out fossil fuels or curb resource extraction. Some promote climate "solutions"—consisting of no more than technological pipe dreams—from which they themselves stand to profit handsomely. Meanwhile, their counterparts in government have been worse than ineffectual. In 2022, the White House and Congress, both under Democratic control, enacted some wholly inadequate policies for climate mitigation with one hand while, with the other, allowing all-time records to be set for oil and gas extraction.[33] Starting in January 2025, the federal government, having come under Republican control, gave top priority to increasing oil and gas extraction even further while barring federal officials or anyone who applies for federal funding from even mentioning climate change. Prospects outside our borders are no better. The annual UN global climate summits (most well-known for the 2015 Paris Agreement) have, in recent years, taken on the atmosphere of an oil-and-gas-industry convention. The exploited nations of the Global South, those most threatened by ecological breakdown, are allotted time to speak at the summits, but their demands to receive compensation for the death and destruction they have suffered because of the North's greenhouse gas emissions are mostly ignored.

Why am I even suggesting a phaseout of fossil fuels when we know it will face such formidable economic and political forces? And why is the degrowth movement urging the kind of societal transformation that Western governments will attack with everything in their power? I'll answer those questions with another: Now that we're caught in a global ecological emergency, what alternative do we have? The hard right has a suicidal urge to further ramp up petroleum drilling and resource extraction, while centrist and liberal figures in government and industry remain devoted to business-friendly technological "solutions," some of which (like electric vehicles) are simply inadequate, while others (like artificial intelligence and geoengineering) could send us hurtling down a path toward the end of civilization. Such false "solutions" have been thoroughly debunked.[34] But widespread fixation on technological half measures and suicidal schemes remains. To break through that fixation and quash the idea that profits and GDP growth must never be jeopardized, we must also argue for entirely plausible scenarios in which societies break free of fossil fuels and follow a path that's not hostile to life on Earth. To sum up, there is an urgent need not only to convince people that global ecological degradation must be reversed, but also to make a convincing case that there are feasible paths toward achieving that goal. One of degrowth advocacy's important roles is to provide a plausible vision of how those paths can lead to future societies that achieve ecological sustainability, social justice and equity, and a better quality of life for all. In the following chapters, I'll show with specificity some of the ways in which such future societies can be far preferable to today's society.

HAPPY SUBTRACTIONS

The dramatic ecological transformation of society that's needed to keep the Earth livable for future generations would also yield some immediate environmental benefits. The 2020 anthropause illustrates some of those advantages: thriving biodiversity, reduced pollution, and a reduction in greenhouse gas emissions. Otherwise, of course, life in the early Covid-19 pandemic was a lousy example for the future. For many, it was a time of increased deprivation, hunger, alienation, and labor exploitation.

The most affluent 1 percent of households captured most of the global economic wealth generated between 2020 and 2022 as

Covid-19 raged; indeed, their share was nearly twice as large as the total that trickled down to the other 99 percent. The 1 percent also produced disproportionately large quantities of greenhouse gas emissions. But they're only 1 percent. It's the wealthiest 25 percent or so who account for the bulk of global emissions.[35] That includes most but not all of us in rich countries. As we'll see in the following chapters, economically stressed and racialized communities within the United States are not only responsible for less environmental damage than others; they are also most harshly impacted—not only by that damage but also by the political, economic, and technological forces that cause the damage.

The chapters that follow focus largely on the United States. Our economy is among those that cause the greatest damage to the earth's living systems. If it continues to expand, it is doomed to fail eventually, causing widespread collapse of the ecosystem and society as we know it. Degrowth, therefore, is rational and imperative. A frequent tendency among those imagining material restraint in affluent societies is to dwell on the need for sacrifice and focus on the desirable elements of our lives that we would be required to give up. Sacrifice is indeed necessary, but it's not the subject of this book. I instead direct attention to some of the pernicious features of twenty-first-century American life that people at various points up and down the economic pyramid can gleefully leave behind as we and future generations reduce the quantities of energy and material resources flowing through society. In other words, I explore what Parrique's "degrowth as emancipation" might look like in everyday life.

I begin with some of the mundane problems that confront us up close in our homes and communities: harms, nuisances, and threats that will recede from our daily lives in a society that abides by ecological limits. But then I zoom out, chapter by chapter, to more systemic improvements that will have more and more transformative impacts on the US economy and society. After home life, I turn to transportation and how much better our lives will be without the environmentally catastrophic support system that's required to keep a quarter-billion personal vehicles at the center of American life. Next, I examine the array of social and ecological ills spun off by the agriculture and food industries, looking at how we can transcend those ills and improve the quality of rural, suburban, and urban life. Finally, I examine the ways in which advocates for degrowth envision a transformation of work life, both on and off the job. In a future economy that has intentionally reduced material production, begun meeting human needs rather than generating profit, and discarded the many notorious hardships and absurdities of the US workday, we can eliminate those types of jobs that we love to hate.

Most people who are already being clobbered by the impacts of ecological breakdown, from the small island nations of the South Pacific, to East Africa, to the shores of the Arctic Ocean, to the charred forest lands of North America, already know very well that the affluent world needs to rein itself in, and quickly. Nor do racialized US communities, targeted over and over with the worst environmental consequences of capitalism, need to be convinced that the privileged must be constrained. But among many people who have known at least adequate material, environmental, and economic security, there may be a widespread worry that if that system is replaced with one that is just and sustainable, much of their security will be lost. It will indeed be a very different country and world, but if we all are to have a livable future, we must step into it. For encouragement, we can look around us at the system we're leaving behind, recognize the calamities it's causing, and imagine a better life without it.

Degrowth will relegate to the past all sorts of everyday problems that impact us collectively as well as personally, problems that can't be banished solely by individual behavioral changes (although they can help). That banishment requires collective mobilization, both locally and globally. And it must account for the well-established fact that the multiplying impacts of climate chaos, land abuse, and pollution are distributed lopsidedly among households and communities, both locally and globally. In line with degrowth principles, this mobilization must ensure that the loss of economic dominance and the burden of reining in the overproduction that's causing ecological breakdown must fall on people, communities, and businesses that have privileged status economically, racially, ethnically, and geographically. In contrast, both the avoidance of environmental calamity and the subtraction from everyday life of the myriad ills spun off by the current growth economy will be enjoyed by all.

In discussing those "happy subtractions," I'll focus on those that we in the Global North, especially the United States, can anticipate. But our reining in of resource extraction and economic expansion will also help end the exploitation and environmental degradation to which we've always subjected the Global South in our quest for perpetual economic growth.

Either for fear of losing their audience or because of their own misplaced faith in technology, environmental activists and opinion leaders tend to steer clear of talk about limits, restraint, sacrifice, and the need to reject material overabundance. Could a focus on happy subtractions help make messaging about the need for ecological limitations more appealing? What if the movement adopts persuasive new talking points about the outrages, annoyances, and grave dangers that plague us and our communities right now but won't be following us into a future that keeps within ecologically necessary limits?

The pursuit of degrowth will honor the living world around us with a new anthropause, one that eliminates our destructive interactions with the ecosphere while fostering healthy ones. In doing so, we can free our own species from the burdens that capitalism and economic growth impose on us every day.

II.

To Ensure Domestic Tranquility

If a society that's living within firm ecological limits is to ensure a decent, fulfilling life for everyone, production of essential goods will need to take priority over other uses of resources. That will require ceasing production of many items that have become everyday features of life in the affluent world over the past century, especially ones that consume a lot of energy and produce harmful waste. Many will mourn the loss of some of these suddenly obsolete artifacts. But I maintain that pivoting away from production and consumption of nonessential goods will make our lives better in many ways and that we will benefit both individually and collectively.

There's probably not a clean split between elements of consumerism that we'll miss and those we won't. For any specific item, not everyone may agree that its exit from our lives will be cause for celebration, at least not at first glance. But I'll try to make a solid, evidence-based case that we will see net benefits when the pursuit of growth is replaced with something more meaningful, satisfying, and beautiful.

Eighty percent of Americans reside in urban and suburban areas, spending most of their time in human-made environments.[36] Companies market many products that do the job for which they're designed but also degrade the quality of life for their owners and most others. I'd like to ease into this gallery of net negatives by starting small—or rather, seemingly small—with that all too familiar neighborhood bully known as the leaf blower.[37]

THE ANSWER, MY FRIEND, ISN'T BLOWIN'

With energy conservation a top priority, a degrowing world will sound a lot better than the one that now surrounds us. We'll hear far more birdcalls and far less earsplitting mechanical racket. One of the noise sources that I expect most Americans will be pleased to be done with is the leaf blower. Few machines are less necessary and more obnoxious. Leaf-raking is a cheap, healthful, and, most of all, quiet activity. The common gas-powered leaf blower is a costly energy hog, imperils the operator's health with toxic fumes, and threatens the earth's health with its carbon dioxide emissions. It doesn't even move leaves efficiently. Most notoriously, it's *loud*.

The leaf blower's ear-shredding wail carries through towns and suburbs, and not only in autumn. Throughout the winter, spring, and summer, it's pushing something around—trash, dirt, grass clippings, leaves, snow—on sidewalks, driveways, and decks. It's used to scour parking lots around businesses, schools, and apartment complexes. Compared to rakes and brooms, it saves little time and performs no better. It's often used inappropriately, especially when there's only a scattering of leaves. In that situation, it would probably be faster to pick the leaves up by hand, or better, just to leave them where they are.

Gas-powered leaf blowers are deafening—literally. In generating wind speeds approaching those of an EF5 tornado, they blast out noise at around 95 to 115 decibels, comparable to what emanates from a chain saw or jet engine. (Federal agencies label noise levels above 85 decibels as hazardous; 95 decibels is twice as loud as 85, while 115 is eight times as loud as 85.) High-powered blowers can cause detectable hearing loss after only fifteen minutes of operation without ear protection, and damage accumulates irreversibly with repeated exposure.[38]

The blower's noise is rich in frequencies of 20 to 500 Hz. This low-pitched growl is the most troublesome component of noise pollution, largely because it carries for long distances and easily passes through walls. Exposure to low-frequency noise (other sources of which include vehicle traffic and wind turbines) is associated with elevated risk for other health problems, including sleep disruption, mental stress, poor cognition, high blood pressure, heart ailments, stroke, cancer, asthma, depression, and immune-system dysfunction. Naturally, these noises are also closely associated with annoyance, which is itself a health hazard.[39]

The noise travels far and loud. Even at eight hundred feet, almost three-football-fields away, tests found that leaf blower racket exceeded the World Health Organization's recommended volume limits.[40] And that's just one piece of equipment. Imagine a homeowner diligently blowing leaves across the lawn in a suburban neighborhood on a Saturday afternoon. Draw a circle of three-tenths-mile diameter with that blower at the center. On that afternoon within that circle, there could very well be other residents, or worse, lawn-care contractors using not only leaf blowers but also lawn mowers, string trimmers, and other gas-powered tools. The cumulative mental and physical health impact on neighbors within that circle can become intolerable.

Noise pollution may often be thought of as something that afflicts only comfortable, well-off homeowners who don't have to worry about more serious threats to their well-being. But according to researchers at the Harvard School of Public Health, the human toll for those exposed to leaf blower noise can be especially high among vulnerable people in our society:

In a densely populated community, people who work from home, people who work night shifts, children, the retired, the elderly, and the sick may be exposed to high level low frequency sound in their homes, apartment complexes, and businesses. Additionally, this area may include people in schools, hospitals, daycare centers, and retirement homes for whom WHO daytime sound standards are 35 dB(A) or less.[41]

That is, more vulnerable people suffer from exposure to noise at much lower volumes than the tolerance limit for the rest of the population. And what about the Americans who have the most intimate, long-term contact with loud lawn equipment: more than one million commercial landscaping employees, whose ears are just a couple of feet away from an industrial-strength leaf blower or other gas-powered equipment for long periods every workday for months at a stretch? People in these generally low-wage, exploitative, seasonal jobs are often recent immigrants or living in otherwise precarious circumstances. They are exposed to dangerous noise levels over long periods of time and are therefore at the highest risk of severe hearing damage from leaf blowers and other power equipment.[42]

The National Institute for Occupational Safety and Health (NIOSH) requires employers to provide hearing protection for workers exposed to loud noise. But not just any earplugs or earmuffs will do. Employers are instructed to provide protection that will tamp sound down to just below 85 decibels but not much lower; stronger protection could be too uncomfortable when worn for long periods and could deafen workers to warnings or emergency instructions. And even if they are provided proper protection, how often do they have no choice but to remove it because of discomfort or a need to communicate? Furthermore, even for companies earnestly trying to comply, NIOSH's procedures for selecting, fitting, and testing of ear protection are highly complex and technical.[43] Given those requirements, the complications involving ear protection, and the varying conditions—acoustic and otherwise—that a team of lawn-care workers faces throughout the year or even the day, there's no guarantee that workers' hearing is continuously protected while on the job.

The machines also generate intense vibrations that can cause operators to suffer a medical condition called "hand-arm vibration syndrome" or "Raynaud's phenomenon." This problem, which can also be caused by connective-tissue diseases or injuries, circulatory-system diseases, exposure to toxins, and some drugs, involves reduced blood flow to the operator's hands, leading to a condition known colloquially as "vibrating white fingers."[44]

Leaf blowers are a clear public nuisance and widely reviled. Cities nationwide have banned or are moving to ban gas-powered models, but blowers are just the tip of the sonic iceberg.[45] Citing US government estimates that the incidence of noise pollution from all sources has been doubling or tripling every three decades, Bianca Bosker wrote for *The Atlantic*,

> Our soundscape has been overpowered by the steady roar of machines: a chorus of cars, planes, trains, pumps, drills, stereos, and turbines; of jackhammers, power saws, chain saws, cellphones, and car alarms, plus generators, ventilators, compressors, street sweepers, helicopters, mowers, and data centers, which are spreading in lockstep with our online obsession and racking up noise complaints along the way.[46]

Furthermore, added Bosker,

> Not everyone bears the brunt of the din equally. Belying its dismissal as a country-club complaint, noise pollution in the US tends to be most severe in poor communities, as well as in neighborhoods with more people of color. A 2017 paper

found that urban noise levels were higher in areas with greater proportions of black, Asian, and Hispanic residents than in predominantly white neighborhoods. Urban areas where a majority of residents live below the poverty line were also subjected to significantly higher levels of nighttime noise, and the study's authors warned that their findings likely underestimated the differences, given that many wealthy homeowners invest in soundproofing. "If you want to access quietness, more and more you have to pay," says Antonella Radicchi, an architect who helps map quiet spaces in cities. Radicchi believes access to quiet havens should be a right for every city dweller, not only the rich, who can afford to escape noise—via spas, silent yoga retreats, lush corporate campuses.[47]

Noise and vibration are not the only dangers posed by leaf blowers. In the United States, annual emissions of nitrogen oxides from blowers, mowers, and other landscaping tools are equivalent to those produced by 30 million gasoline-powered cars. For fine particulate matter, the equivalent rises to 234 million cars, and for carbon dioxide, it's 6.6 million cars. Lawn-care equipment also emits more than 20 million pounds of carcinogenic benzene into the environment. Some portion of the particulate matter and benzene can move straight into the nearest set of lungs: those of a maskless operator.[48]

Because commercial-grade blowers are generally more powerful and are operated for many more hours per year than consumer-grade models, they're responsible for the bulk of national emissions from lawn and garden equipment: 82 percent of fine particulate matter, 77 percent of nitrogen oxides, and 67 percent of volatile organic compounds. Therefore, as with noise, it's the members of the landscaping workforce who are most affected by the toxins.[49]

Finally, the American penchant for total leaf removal has hyperlocal ecological impacts. Leaf litter provides critical habitat for small mammals, insects (including pollinators), and a host of much tinier organisms vital to soil health. Leaves thinly scattered on grassy areas, used as mulch around trees and bushes, or covering the garden in the offseason, support biodiversity and soil health. Allyson Chiu reported for *The Washington Post*, "Experts said taking steps to leave leaves around your yard in the fall is just one part of a larger effort to move away from manicured lawns that are dead zones for many species, providing little to no habitat or nutrients for wildlife and requiring many resources, including fertilizer, to maintain."[50]

Once we've stopped gathering, bagging, and hauling away valuable organic matter in the form of leaves, whether they've been blown or raked, we'll be keeping ten million tons of waste per year out of US landfills. That currently makes up 7 percent of *all* landfill waste, according to Chiu, citing the EPA.[51]

Degrowth advocacy often features the concept of convivial tools and technologies. According to the twentieth-century philosopher Ivan Illich and those who have followed him in this line of thought, a convivial tool is one that its user fully understands and controls, that anyone can use at any time for their purposes, that doesn't worsen exploitation, and that empowers a person to do work that they would be unable to do without a tool.[52] He might have added that it shouldn't impair the well-being of those exposed to it. With the absence of leaf blowers and other decidedly nonconvivial tools, a society centered on degrowth would be much more congenial and just, and certainly less noisy, than today's.

LIGHTS OUT WHEN THE STARS ARE OUT!

Minimizing energy use will require putting artificial lighting, indoors and out, on an energy diet. That will have dramatic effects, among them a return of a nighttime sky that has been largely hidden from much of humanity in recent decades. Today, more than 80 percent of the world's people look up at night to a sky polluted by artificial light. For US and European populations, the share rises to 99 percent. Today, for three out of five Europeans and four out of five North Americans, the Milky Way has completely vanished, except when a widespread power blackout happens to strike on a clear, new-moon night.[53]

Light pollution of the night sky over human-populated areas of North America and Europe has been increasing at the astonishing rate of 10 percent per year. That estimate comes from the Globe at Night project, in which more than fifty thousand nonspecialist volunteers ("citizen scientists") were issued sky maps and agreed to go outdoors on clear nights between 2011 and 2022 and identify the dimmest star they could see with the naked eye. From those data, researchers discerned a 10 percent yearly increase is what they call "skyglow"—a brightening of the night sky caused by refraction of artificial light that emanates from outdoor sources such as streetlamps, outdoor athletic facilities, and illuminated advertising. With the brightening of the night-sky background, increasing numbers of stars and other celestial features have been bleached out of our view.[54]

This rapid nocturnal brightening has resulted in part from a well-intentioned effort to improve energy efficiency and reduce carbon emissions by replacing traditional lighting sources with light-emitting diode (LED) lamps. LEDs shot up from less than 1 percent of new lighting worldwide in 2011 to 47 percent in 2019. By 2022, 66 percent of all new outdoor illumination in the United States was provided by LED lamps. The switch was propelled by LEDs' ability to produce far more light per watt of energy consumed than do traditional light sources. But instead of using LEDs to provide the same amount of light as before but at lower energy costs, companies and municipalities took advantage of the technology's energy efficiency and began bathing parking lots, streets, billboards, sports fields, car dealerships, and other outdoor spaces in blinding illumination.[55]

The sheer increase in brightness accounts for part of the post-2011 spike in skyglow, but it has been further boosted by another property of LED lighting: its richness in short wavelengths at the so-called cool-blue end of the visible spectrum. This blue light is scattered by the atmosphere much more efficiently than longer wavelengths, and a good share of the scatter returns to Earth. Blue light also bounces from streets and parking lots toward the sky more readily than other wavelengths. All of this makes the sky glow brighter and fades out the stars.[56]

Nighttime darkness has become increasingly scarce, even for those who don't live in an urban area. Light from a city of 1.5 million can significantly bleach out the sky at locations fifty miles away. Skyglow from even a town of thirty thousand can extend ten miles. Skyglow is further exacerbated by air pollution and dust.[57]

Within a little more than a decade, excessive LED lighting has all but erased many people's celestial views. Before publication of the Globe at Night visual-observation results in 2022, it was generally believed that the night sky was brightening at a rate of "only" about 2 percent per year, not 10. It turns out that the 2-percent figure was based on data collected by satellites that don't detect the short-wavelength blue light produced by LED lamps. In contrast, the human eye is very sensitive to short wavelengths, especially at night. So, while satellites were blind to skyglow produced by the LED takeover of the lighting market, the Globe at Night citizen scientists watched it happen in real time.[58]

The researchers who analyzed the Globe at Night data put the result into perspective this way: "For an eighteen-year period (such as the duration of a human childhood), this rate of change would produce more than a factor of four increase in sky brightness. A location with 250 visible stars would see that reduce to 100 visible stars over the same period."[59] The good news is that light pollution is quickly reversible in a way that, say, carbon dioxide pollution is not. The very rapidity of the night-sky loss tells us that energy-saving measures to put humans' outdoor environment on a strict dimmer switch could quickly make visible again the innumerable stars that we enjoyed in the past. Further scale-back of outdoor lighting would bring back sights like the Milky Way, meteors, comets, and the Andromeda galaxy that disappeared from most people's view decades ago. As a bonus, the restored darkness could also improve our mental, emotional, and physical health.

Amid mounting ecological and humanitarian crises, the vanishing of the heavens may be regarded by some as a phenomenon of concern only to astronomers and aesthetes. This view badly underestimates the central role the night sky has played in the history of our species and ignores the threats that excessive artificial lighting poses to human health and the living world around us.

Arguing in a 2010 article that the night sky has provided deep scientific and cultural value, as well as recreation, inspiration, and, above all, beauty, the Missouri State University economics professor Terrel Gallaway concluded,

> Indeed, if the continuity of human life is to mean something more than perpetual breeding, then there is a strong case for preserving easy access to a cultural asset that has been a part of art, math, science, and culture for as long as these things have existed. Moreover, the night sky is an inclusive, community-building good. In the absence of light pollution, it is exceptionally non-invidious. Nightscapes are freely available to virtually all people . . . Artificial lighting . . . is waste that prevents hundreds of millions of people from seeing an important part of their shared history, converting a public good into one available only to those fortunate to live under dark skies or with income sufficient to travel to remote locations.[60]

Addressing fellow economists, Gallaway noted that "utility bills do not well reflect what is culturally special about dark skies—the beauty that is the night's true comparative advantage. It is hard to imagine, therefore, a long-term solution to light pollution that does not explicitly recognize the fundamental role natural beauty plays in social welfare."

More broadly, ecological light pollution, defined as "artificial light that alters the natural patterns of light and dark in ecosystems," has far-reaching consequences. Its sources include street lamps, parking lot lights (perhaps most notoriously, those at car dealerships), decorative displays, vehicle headlamps, airports, flares at oil and gas facilities, floodlighting of building facades, security lighting, and countless others.[61]

Light pollution disrupts a wide range of animal behavior and development, including hunting, mating rituals and reproduction, feeding, growth rate, communication, and migration.[62] Excessive lighting is one of many human activities that have severely driven down insect populations. The decline can be seen most vividly in surveys of "bug splatter" on vehicle windshields, which has decreased by two-thirds since 2000.[63] A 2023 review of light-pollution research published in the journal *Science* found that "light at night has been implicated in the recently observed insect die-off, because of decreased reproduction and a fatal attraction to car headlights, streetlights, or other lighting."[64]

Nocturnal insects are most sensitive to light of cool-blue wavelengths. Experiments show that they are more strongly attracted to blue-rich LED street lamps than to red-rich light from the old high-pressure sodium lamps, and that puts them at even greater risk.[65] In a tragically ironic twist, the firefly, that much-beloved summertime showoff, is among the many insects threatened by light pollution. This family of beetles, many species of which have evolved the ability to generate light through their own metabolism, is now seeing their breeding success and larval dispersal sabotaged by industrially produced light.

An estimated 30 percent of North American firefly species are either vulnerable to or in danger of extinction. They're threatened by overall habitat degradation, but excessive lighting plays an especially important role, by disrupting reproduction. For example, females and males of those familiar species known as flashing fireflies or lightning bugs flicker precisely timed codes to attract possible mates. Too much skyglow or light from ground sources obscures their signals, and as a result, they produce fewer offspring.[66]

Fireflies are valuable not only culturally but also as integral parts of the ecosystems they inhabit as larvae and adults. Their ecological importance is generally undervalued. To make matters worse, some lighting engineers are layering even more cruel irony onto the firefly's predicament by applying what they've learned from studying the insect to further increase the brightness of LED lighting. Korean researchers have designed a new type of LED lens by emulating the lens-like layer that covers the firefly's bioluminescent abdomen. They report, "The biologically inspired LED lens . . . substantially increases light transmission over visible ranges . . . This biological inspiration can offer new opportunities for increasing the light extraction efficiency of high-power LED packages." Millions of years of firefly evolution produced a highly efficient lens that's now being copied by humans in a way that could hasten the extinction of the insects that evolved it.[67]

All classes of animals, including insects, fishes and other sea creatures, amphibians, reptiles, and mammals, are disrupted by artificial nighttime light. Bats—like lightning bugs, nocturnal fliers suffering dangerous population decline—are among the mammalian species most vulnerable to light at night, especially where facades of churches, old historical buildings, or other popular bat habitats are illuminated through the night. Wallabies, squirrels, voles, lemurs, roes, geckos, salamanders, and lizards, among many other species, have been shown to suffer ill effects. A well-known case involves loggerhead turtles, who lay their eggs on sea beaches. Normally, new hatchlings head straight for the surf, which they're able locate under very dim light. But on commercialized beachfronts with lots of light pollution, the young turtles become disoriented and often fail to reach the water.[68]

Like humans, some other animal species use star patterns for orientation, especially during migration; not surprisingly, the recent, rapid increase in skyglow is having disproportionate impacts on them.[69] Light pollution causes mood changes in hamsters and mice, and that knowledge, experts say, could contribute to the study of depressive disorders in humans.[70]

Human problems with nighttime light pollution, like those of our fellow animal species, are often connected to disruption of our circadian rhythm. Both hormonal secretions and the brain's neural circuits are disrupted. Crucially, interrupting the darkness of nighttime suppresses the production and circulation of melatonin, the hormone that regulates sleep patterns. Impacts on health can be severe: gastrointestinal disorders, diabetes, cardiovascular disease, and cancers of the breast, lung, colon, bladder, pancreas, and prostate. Other risks include skin disorders, obesity, Alzheimer's disease, depression, suicide, and even Covid-19 infection.[71]

By putting civilization on a dimmer switch, we would not only conserve energy, reduce greenhouse gas emissions, bring back the night sky, save the lightning bugs, and ameliorate threats to our health—we could also produce big advances in environmental justice. Across the United States, light pollution is more severe in neighborhoods where a larger proportion of the population is Black, Latino, or Asian. The disparity in nighttime lighting is especially prominent in urban cores but significant in suburbs as well. The sources of this avoidable health hazard are similar to those of other environmental injustices. Industrial and commercial facilities, many of which are brightly lit at night, have long been built in or near low-income, racialized communities that are targeted because they don't have the economic or political power to keep them out.[72]

The problem is worsening. As dark night skies have become more highly valued, residents of privileged neighborhoods have fought harder than ever to protect their property values by making sure bright light sources continue to be sited in parts of town that are already over-lit. Crucially, the elimination of dark spaces is also a central element in stereotyping residents of Black and Latino neighborhoods as criminals.[73]

That last point is rooted not only in systemic racism but also in wrongheaded beliefs about crime. Research has found scant evidence that rates of crime or traffic accidents rise when street-lighting is reduced. Nevertheless, affluent residents of urban and suburban areas insist that light pollution is necessary for crime prevention in less affluent neighborhoods, especially mixed residential-commercial areas.[74] Racist notions about crime thereby account for much of the greater exposure to artificial light at night in Asian, Latino, and Black neighborhoods than in mostly white neighborhoods. Even when comparisons are made only *within* highly light-polluted areas, Asian, Latino, and Black residents suffer twice as much exposure as their white counterparts.[75]

I think it's fair to say that with degrowth and judicious allocation of limited electric power, excessive illumination of the outdoors will fade into the past. Electricity would be used to meet societal needs; blindingly lit-up car lots and LED billboards will have no place in that future. Improvements in energy efficiency will be seen as an opportunity to conserve energy, not produce even more harm. Nighttime outdoor lighting will be used only where and when necessary and otherwise treated as a pollutant. Meanwhile, to quote Terrel Gallaway once more, a restored view of the heavens in their full splendor will have social value as "an inclusive, community-building good," with "nightscapes . . . freely available to virtually all people."

NOT COOL

I've been writing critically about air conditioning for the past fifteen years, so I naturally wanted to include it among the technologies on which we'll need to become less dependent in an energy-restricted future.[76] But in a rapidly warming world, we AC critics need to fine-tune our argument. Even in a degrowth society, it should be stressed, we will still need enough AC capacity to preserve health and life during the heat waves that will continue to grow longer, hotter, and more frequent. Also, most people will not want to give up air conditioning. Love of its cool breeze is a deeply held sentiment for many, so we will need to highlight the many benefits that natural ventilation provides when there's no heat wave in progress.

There remains a critical need to cut back on the use of refrigerative cooling. Booming demand for air conditioning will keep pushing global greenhouse gas emissions higher in the coming decades, marginal improvements in energy efficiency and refrigerant technology notwithstanding.[77] Minimizing air conditioning's climate impact, therefore, will require less use during garden-variety hot weather, which most of us experience for many more days per year than we do extreme heat waves.

The benefits of deeply reduced air conditioning use would be legion. Running it less would boost our bodies' physiological heat acclimatization while also increasing our mental tolerance for heat. The "adaptive model of comfort," a result of extensive research, shows that the temperature range in which we feel comfortable isn't fixed but slides higher or lower depending on the temperatures our bodies have experienced in recent days or weeks. If we're amply exposed to the warming weather of spring and summer, our mindset acclimatizes along with our circulatory system and sweat glands.[78]

Air conditioning gets well-deserved credit for lowering the risk of heat stress and mortality when temperatures are dangerously high. But those of us who are not elderly or otherwise vulnerable to heat might improve our health during more routine hot weather by keeping the AC system off and windows open. There's ample evidence that living and working in continuously cooled environments can increase risks of vitamin D deficiency, obesity, and hormonal disruption. Studies conducted in France, the United Kingdom, Brazil, Denmark, California, and elsewhere found that people who worked in naturally ventilated spaces had fewer health problems than those who spent all day in air-conditioned workplaces. Some of those differences may have been connected to the buildup of air pollutants and pathogens in well-sealed, energy-efficient environments with inadequate fresh-air turnover.[79]

In a future with much less air conditioning, people would spend much more time outdoors. Being out amid fresh air and vegetation can also confer better physical and mental health prospects. *The Lancet* reports,

> There is increasing evidence that access to urban green spaces provides benefits to human physical and mental health. This includes reducing exposures to air pollution, relieving stress, and increasing social interaction and physical activity, with overall improved general health outcomes and lower mortality risk.[80]

We're in dire need of more social interaction, another benefit that could come with lower AC reliance and a resultant shift toward more outdoor life. To see the need for this, try revisiting a neighborhood where, in summers past, residents would sit and chat in the shade while kids romped from yard to yard or stoop to stoop. Today, vehicles parked in driveways or along curbs may be the only clue that human beings are present in the area during the summer. On a recent, beautiful eighty-nine-degree early evening,

I took a leisurely five-mile bike ride through shady residential neighborhoods near my home. Scanning left and right, I could not spot a single house with its windows open. Come to think of it, I don't recall seeing any people either.

Air conditioning doesn't bear total responsibility for the near demise of outdoor social life. That loss is also due in part to the way digital entertainment and communications hold our society in their grip. However, as I wrote in 2010, "Full enjoyment of a jumbo-screen TV, a PS3, a DVD, a PC, an SUV, or an RV calls for AC."[81] One by one, the products in that list have faded into obsolescence, but every summer, digital devices still work in tandem with air conditioning to lure people into their homes and vehicles and keep them there. Leaving *over*use of air conditioning behind would help restore our cultural capacity for living more of life outdoors, collectively.

ADIEU TO ADS

Noise and light pollution can take many other forms and may be at their most damaging and annoying when working as a tag team. I submit that such is the case with advertising via electronic media. Ridding society of advertising's assault on our senses— full of sound and fiery light, illuminating nothing—would bring welcome relief. In societies that make the collective decision to produce much smaller quantities of nonessential material goods, advertising will be pointless anyway. And surveys suggest that few would lament its absence. Even many of those who work in the advertising industry today won't miss it when it's gone.

A capitalist economy requires growth for its very survival, and continual growth depends on continual increases in consumption. To keep consumption high in a society where most people's essential needs are fully satisfied, potent stimuli are required. Despite being widely regarded as a public nuisance, advertising has been used for more than a century to keep this Ponzi scheme churning. In the process, the ad industry has become more imposing than ever.

Among the many unwelcome intrusions into our daily life, ads bulk large. When asked if they liked advertising, only 11 percent of UK respondents said yes, even though the question was slanted by the inclusion of a suggestion that ads "can be enjoyable." In a US poll asking respondents to rate people in twenty-one different professions on their ethics and honesty, advertising professionals ranked fourth from the bottom, with a 10 percent approval. Whom did they edge out, you may ask? Telemarketers, car sales-people, and politicians.[82]

In a 2019 *The New York Times* article titled, "The Advertising Industry Has a Problem: People Hate Ads," reporter Tiffany Hsu led with this: "In the predigital days, advertising agencies were ruled by swaggering creative directors who gorged on lavish client con-tracts and sometimes created campaigns that set the cultural agenda and captivated the public. Nearly every piece of that equation has changed." Now, reported Hsu, Americans are exposed to five times as many advertisements per day than in the 1970s. And even worse than the increased volume is the nature of ads. The same digital industries that have taken a sledgehammer to journalism have spawned a new species of advertising that's even pushier and more annoying than ads were in the age dominated by print, radio, and television. Without ad-blocking and anti-tracker software, simply catching up with the day's news online can be like walking through a crowded bazaar teeming with hawkers and pickpockets.[83]

Writing for the trade publication *Marketing Week* in 2019, branding consultant Mark Ritson quipped, "At some point at every marketing conference you've ever been to, someone gets up on stage and declares (drum roll, please): 'People don't hate adver-tising. They hate bad advertising.'" If you believe that, Ritson advised readers, you're fooling yourself. If you're an advertiser, the general attitude toward your work product is much simpler. In his words, "People hate it. Good ads. Bad ads. All ads."

In 2024, as the ad industry was scaling new heights of vexatiousness, the writer Kate Lindsay described the extent to which marketers have party-crashed every nook and cranny of the digital world, cluttering screens and speakers with tacky intrusions that are *designed* to be annoying, because that rouses our attention. Meanwhile, we give our personal data away to the corporate internet for free, and they use that information to tailor individually targeted ads. Lindsay reminds us, "The more data a platform collects from your online browsing habits, the more powerful it is, and Meta knows who we are, Google knows what we're searching for, and Amazon knows what we're already buying." Now, she adds, businesses of every size are getting into the game:

> A big part of why the internet has become an adpocalypse is that this kind of targeted advertising is no longer reserved for the tech giants. In recent years, diet apps, fitness apps, period-tracking apps, transportation apps, dating apps, food-delivery apps, and basically every other kind of app realized they have valuable personal information that we, by agreeing to their terms and conditions, have allowed them to access. Now they're monetizing it.[84]

In a degrowing world that has no need or use for advertising, we would be spared this endless pestering and surveillance. We'd also get relief from what may be the most dangerous ecological threat posed by advertising: its effectiveness.

In a 2021 article, four professors of business and advertising reviewed the academic literature, seeking to answer the question, "Does advertising, in fact, help or hinder consumer well-being?" They noted that "At its best, this critical form of communication taps into the hearts, minds, and ambitions of consumers to enhance their quality of life," but at its worst, advertising can "engineer frivolous needs, wants and desires" and have "psychological influence that increases purchase likelihood, changes consumers tastes, and may even change how consumers think about themselves."[85] This wasn't exactly news; it was well-known even before Vance Packard's enormously influential book *The Hidden Persuaders*, published in 1957, shoved public relations and advertising into a harsh public spotlight.[86] Today, however, advertising is at its worst far more often than at its best.

In the introduction to a fiftieth-anniversary edition of Packard's bestseller, Mark Crispin Miller wrote,

> As this book revealed, the industry was trying to hone its probing apparatus to so fine a point that the staunchest skeptic would not see it coming, and its long needles would now reach so deeply into him, without his even knowing it, that he might buy just as the apparatus meant him to, while quite convinced that such "decisions" were his own.[87]

As a result, even our exasperation with ads fails to undercut their influence. Annoying online ads are better remembered than comforting ones.[88] When ads succeed in evoking desire and materialism, they undermine their human targets' quality of life in several ways. People who buy products that, according to ads, will improve their lives and then see no such result often take a blow to their sense of well-being. Ads can also lead to neglect of social relationships and excessive debt. In stark contrast, people who strongly value intrinsic sources of happiness and have low-material desires—ideas consistent with degrowth—see their quality of life improve.[89]

But wait—as the old TV ads for Ronco gadgets would invariably inform the viewer—there's more![90] Decreased material footprints and the consequent demise of the advertising industry will not only improve our collective quality of life; the whole

Earth will benefit. Through its own energy consumption, the US online ad industry is directly responsible for more than sixty million tons of carbon dioxide emissions annually. Then there are the emissions for which it's indirectly responsible, that is, when ads spur excess consumption that would not have occurred otherwise. Through its stimulation of materialism, consumption, and investment—and thereby through economic growth, resource consumption, and ecological disruption—the ad industry further amplifies its destruction. Corporations compound the damage with phony eco-advertising. Ads in various shades of green, scenes of the outdoors, terms like "smart," "fresh," and "eco-," all appear to be effective even when the product has no redeemable properties, environmentally speaking.[91] Such greenwashing lures well-intentioned people into material overconsumption.

Advertising also provides financial underpinnings for the still-growing online world of social media, video streaming, retail goods, and much more—and that world has a massive carbon footprint. Research recently published in the journal *Nature Communications* shows that the networks and devices that distribute and deliver online content account for huge quantities of carbon dioxide emissions and insupportable extraction and use of minerals. Continuing as it is today, but surrounded by a world that's on the path to preventing 1.5 degrees of global warming, the online world alone would capture more than 40 percent of the global-carbon budget. And it would hog 55 percent of the mineral tonnage that can be extracted without breaching other crucial planetary limits.[92]

With deep cuts in resource use, and with material production erasing any perceived need for advertising, people in a degrowth future would not be swamped by a cornucopia of manufactured goods. With everyone, including poor and low-income people, gaining assured access to the material goods required for a good quality of life, and with the manufacture of desire fading into the past, we will find our lives far richer in nonmaterial ways.[93]

MATERIAL WORLDS

Three decades ago, the photojournalist Peter Menzel's extraordinary book *Material World: A Global Family Portrait* was published. For the book, Menzel and sixteen other photographers fanned out across the world to spend time with one "statistically average" family in each of thirty different countries. The members of each household agreed to pose for a group photo outside their home with all their worldly belongings arrayed around them—everything that could be moved. Revealing as it did the enormous disparities in wealth and household consumption across our planet in unforgettable imagery, *Material World* remains, to me, the world's greatest coffee-table book.[94]

It's impossible to capture the full scope of the photos in words, but I'll just mention the features that have always been most striking to me. And because members of each family were interviewed about their most valued possession(s) and their wishes for the future, I'll include some of their responses, which were almost as interesting to me as the photographs.

The Natomo family in Mali posed on their earthen home's flat rooftop with a spare collection of items that included an array of farm and cooking tools and an enormous mortar and pestle, apparently of stone, and a bicycle, their most valued possession. The family was hoping someday to own an irrigation system and an enclosed garden, and upgrade to a motorcycle.

In Ethiopia, the Getu family posed with a team of oxen (their most-valued possessions), a horse, a sheep, a chicken, and not much else. They wished for more animals, better seed stocks, some farm implements, and "peace in the area and the world."

The Delfoart family of Haiti also had an array of livestock: a horse, a chicken, a donkey, and a goat. The parents said they had "no possession of value" and apparently were mum on their wishes for the future. Their typical diet included cassava, potatoes, smoked fish, and coffee for breakfast; potatoes for lunch; and, they said, "nothing" for supper.

Mongolia's Regzen family were proud owners of the book's most visually striking home: a circular, wood-framed, felt-and-canvas-covered tent called a *ger*, the traditional home in Mongolian society (and more commonly known in the West by the Russian word *yurt*). They posed amid attractive furnishings. To the mother, the family's most prized possession was a Buddha statue, while the father voted for the TV. Their common wish was for a "permanent house for all seasons."

In Cuba, the Costa family's possessions, including four bicycles, three TVs, and plenty of furniture, filled the narrow street in front of their home. They made no mention of valued possessions but aspired to own a car and a video game player.

The Kalnazarov family of Uzbekistan had the most unusual collection of material assets. It consisted mostly of several collections of lovely carpets and blankets, each stacked several feet high. According to the sons, though, the family's most prized possessions were bicycles. They were wishing for a TV, radio, VCR, and car.

In Italy, the Pelligrini family displayed an impressive array of furnishings. The father's most valued item was the mandolin with which he posed in the photo, while the mother cited the antique-doll collection displayed behind her. Like the Kalnazarovs, they wished for a VCR (this was the early 1990s, after all), but their more ambitious hope was to one day have a farm.

The front view of the Skeen family's home in Pearland, Texas, was dominated by an open two-car garage containing a pickup truck and van. Their possessions filled the front yard and driveway, spilling out into the turnaround of their cul-de-sac. The mother held a large open Bible (their most valued possession), while the father held a cowboy hat over his heart. Their wishes for the future included tools, new carpeting, and a camping trailer.

Upon casual observation, the material trove owned by Germany's Pfitzner family appeared to eclipse even the Skeens's. Nevertheless, their most valued possessions were humble: for the parents, a basket of family memorabilia, and for the two young sons, a pocketknife and a toy truck. Their wishes for the future, however, were expansive: a new refrigerator, a house in the country, and, finally, a cleaner environment.

Most of these and the other twenty-one families, ranging from impoverished to affluent by global standards, quite naturally valued material goods that had utility for them, while many also assigned high priority to nonmaterialist desires, citing their memorabilia and religious artifacts, family, nature, gardens, leisure time, world peace, and so forth. Another general trend was even clearer: The more material property a family had, the more they wanted.

DECENT LIVING STANDARDS

The material excess of the Global North didn't arise from individual greed or acquisitiveness. It was created by capitalism and the growth that is its lifeblood. Businesses produce far more goods than are required to meet a society's needs. They're compelled, therefore, to persuade people to spend money on those goods not only through manipulative advertising but also through planned obsolescence, the release of "new-and-improved" versions of the same products, and other tricks of the trade. Writing in the early twentieth century, at a time when such ploys were being developed and refined, the legendary economist and social critic Thorstein Veblen saw them for what they were. Echoing Plato's famous aphorism (but flipping it), he once quipped, "Invention is the mother of necessity."[95]

But what do we really need? In his much-cited mapping of human requirements, the late Chilean economist Manfred Max-Neef wrote that universal human needs include subsistence, protection, understanding, participation, affection, idleness, creation, identity, and freedom. He added that the need for subsistence must be satisfied if the other eight universal needs are to be met. Doing that requires adequate availability of what Max-Neef termed "satisfiers" of the need for subsistence, which he identified as shelter, energy sources, clothing, food, water, health care, communication, and transportation.

Ample research has been devoted to quantifying those satisfiers by coming up with global estimates for minimum per-household material requirements for a good quality of life. One set of widely cited estimates—technically referred to as "decent living standards"—includes the following:[96]

o   Housing with solid roof and walls—of brick, wood, concrete, or similar construction and having a minimum floorspace of 430 square feet for a four-member household (in contrast with the US median new-home floorspace of 2,500 square feet).[97]

o   Electrical lighting; electric or gas cooking; refrigerator with a volume of at least four cubic feet; heating or cooling equipment if necessary; household electric supply.

o   Adequate, reliable water supply, of minimum thirteen gallons per person, per day, from an accessible water source (in contrast with US-household water consumption of eighty-two gallons per person); and an indoor toilet.

o   Local infrastructure for electricity, water, and sanitation.

o   Sufficient clothing to achieve basic comfort in the prevailing climate or a certain amount of clothing with adequate materials catered to a local climate; a minimum number of shared washing machines per one thousand people.

o   One phone and one television or computer monitor per household, supported by accessible communication infrastructure.

o   Minimum public gathering space per one thousand residents, with adequate—but not excessive, I would add—lighting at night.

Clearly, between the consumption patterns of middle- and upper-class American life and the minimum standards for a decent life there lies ample room for degrowth in use of materials and energy. But the right social conditions need to be created. One analysis of the world's countries showed that an energy supply of 1,300 watts per capita is sufficient to support decent living standards and, crucially, that increasing energy availability beyond 2,000 watts per person does not bring additional improvement in measures of human well-being. With US energy consumption exceeding 9,000 watts per person, it's clear that there's a huge surplus just waiting to be eliminated in the process of degrowth.[98]

At the other end of the spectrum, almost half of the world's nations do not even have access to the minimum 1,300 watts per capita worth of energy that's required to provide decent living standards. Many people in those and other nations also don't have adequate housing or local infrastructure. If they are to provide those elements of a decent life, those nations must gain increased access to fossil fuels and mineral resources. To make that happen, the Global North will need, at a minimum, to cancel the South's foreign debts and provide financial aid. The South's higher fuel and resource use, necessary for well-being, will increase its climate and ecological impact, making it crucial that we in the North make even deeper cuts in our much larger energy and resource consumption and reduce our production of wastes, including greenhouse gases, even faster.

Adequate supplies of energy and materials are necessary but not sufficient to guarantee that human needs will be satisfied. Therefore, some of the same authors who promulgated decent living standards have also proposed a set of socioeconomic conditions that can help a country achieve a decent life for all with modest or low-energy consumption. They include a high degree of income equality, high quality of public services, and, eventually, less dependence on resource extraction.[99] All of that is eminently consistent with degrowth.

Material World, Max-Neef, and research on decent living standards all point to the conclusion that tens of millions of US households could significantly reduce their material and ecological footprints without doing any harm to their own well-being or quality of life. There's ample evidence, in fact, that the growing overhang of manufactured and constructed products causes a great deal of stress, anxiety, unhappiness, and conflict. Saying goodbye to that excess, while simultaneously ensuring sufficiency for all, would relieve society of two great burdens at once.

DON'T COPY, BE HAPPY

Seeking to acquire ever greater quantities of material possessions can end only in failure; for one thing, studies show that materialism is associated with a poorer sense of well-being. Money improves life satisfaction when it's needed to stay out of poverty, but it provides decreasing marginal returns when acquisitiveness rises with increasing wealth and income.[100] This tendency has been observed for many decades. In his 1909 book *The Theory of the Leisure Class*, Thorstein Veblen ventured an explanation, one that's widely accepted today:

> The standard of expenditure which commonly guides our efforts is not the average, ordinary expenditure already achieved; it is an ideal of consumption that lies just beyond our reach, or to reach which requires some strain. The motive is emulation . . . each class envies and emulates the class next above it in the social scale, while it rarely compares itself with those below or those who are considerably in advance.[101]

We see emulation at work in the notorious supersizing of American homes in recent decades and in the accumulation of consumer goods. The median US family home has floorspace of about 1,600 square feet. That's 60 to 100 percent more space than the median house in most comparable countries. These days, the median size of newly built houses exceeds 2,000 square feet. Joe Pinsker, a former staff writer for *The Atlantic*, has listed a slew of factors that pushed house size so high: minimum square footage requirements in zoning laws; reduced construction costs; homebuilders' marketing tactics; the growth of commuting by car from low-density suburbs with large lots; and perhaps most importantly, the prevalent view that a house serves first and foremost to preserve and augment its owners' net worth and only secondarily to provide shelter.[102]

With the number of occupants per US home at an all-time low, square footage per person exceeds that of an entire median-size house in Europe. That leaves lots of floorspace unoccupied most of the time, and rooms devoid of humans tend to fill up with things. The book *Life at Home in the Twenty-First Century: 32 Families Open Their Doors*, published in 2017, documents the kind of material abundance on display in that Texas driveway in

Material World, but from an indoor viewpoint. None of the families featured in the book had more than five members, but its photographs are intensely claustrophobic, thanks to the material abundance that's crammed into each image.[103]

A survey of home-organizing professionals found, not unexpectedly, that larger homes attract more stuff: "People's explanations for having too many things fall into two categories, namely: (1) We buy more than before because we have the money; and (2) we retain more unwanted things because we have the space and lack the time to sort, organize, and dispose of unwanted things."[104] Here we have a feedback loop. People with plenty of disposable income buy the biggest house they can afford, and, over time, they purchase enough consumer goods to fill all available space. Stressed out by dealing with the mountain of material possessions but not wanting to part with them, they may go shopping for an even bigger house. They'll soon set about filling that house with stuff as well, because they once again "have the space." (The feedback loop does, however, have a pressure-release valve: more than fifty thousand self-storage rental facilities in the United States, offering sixteen billion more cubic feet of space ready to be filled with excess stuff. And business, apparently, is booming.)[105]

In a degrowth-style economy, we would compensate for reductions in aggregate material wealth by looking to solidarity and mutual aid. And fortunately, in contrast to material consumption, there are to be no Veblen effects when it comes to the sharing economy. In fact, while our neighbors' greater wealth may lead us to become less satisfied with our own economic position and want to copy them, seeing those around us build better social relationships makes us more, not less, satisfied with our own personal networks.[106]

When people set aside materialistic goals and values, research says, they see improvements in self-esteem and well-being. Conversely, a focus on materialism can drive out social relationships, and that can bring on feelings of loneliness. Worse, according to Rik Pieters, a professor of marketing at Tilburg University, "sustained loneliness can raise social anxiety, pessimistic social expectations, hostility toward others, and active social distancing. This further frustrates the basic need for relatedness and prompts new material compensation and substitution to cope with it."[107] Materialism is incompatible with collective values, and this contradiction damages psychological well-being. In contrast, there's a positive association between "pro-environmental behavior" and life satisfaction.[108]

Degrowth would bring collective life satisfaction in many ways. For one thing, we could say goodbye to those irrational aspects of current home and community life that bring plagues of air, light, and noise pollution, sleep disorders, labor exploitation, and health crises, while depriving us of the night sky, biodiversity, and possibly our hearing. Putting an end to both overconsumption and poverty-induced underconsumption would cure materialism, foster solidarity, and thereby dispel feelings of isolation. Homes and communities would no longer be reduced to the status of marketplaces and infested with round-the-clock advertising. With all that, we'd be freer to come up with and practice ways of life that will be necessary in what is sure to be a more unsettled future, both ecologically and politically.[109]

Getting Off the Road
to Nowhere

Writing about motor vehicles in the early 2000s, André Gorz, the coiner of *décroissance*, harked back a century into the past:

> When the car was invented, it was intended to procure an entirely novel privilege for a number of very rich bourgeois: the privilege of traveling much faster than everybody else. Up to that point, no one had thought of this . . . the lord's carriage went no faster than the peasant's cart, and trains carried everyone at the same speed . . . The car . . . extended class distinction to speed and transport.[110]

By the time Gorz wrote those words, the rich had lost the distinction of moving faster than everyone else. Motor vehicles had become thoroughly democratized, at least in Europe and North America, with hundreds of millions of them clogging streets and highways. Once again, everyone was moving as slowly as everyone else. The car had become a plague on urban life, and driving one had become a chore. Gorz observed:

> The attractions and advantages of cars, like villas on the Riviera, depend on their being beyond the reach of the masses.

That is because, both in design and original purpose, the car is a luxury good. And . . . if everyone has luxury, then no one has the advantages of luxury. Quite the opposite, in fact. Everyone frustrates and dispossesses others.[III]

Today, alas, capitalism's inherent dependence on economic growth requires affluent economies to continue ignoring Gorz's unimpeachable logic. Once they existed, motor vehicles proved too profitable *not* to be mass-produced and mass-marketed; invention proved once again to be the mother of necessity. And wherever they swarmed urban areas, cars degraded everyone's quality of life.

City dwellers moved to the suburbs to escape the ills imposed by the car only to become long-distance commuters wholly dependent on the car and subject to the worst that the age of personal mobility had to offer. Today, people who commute by car suffer more mental stress, sleep disorders, and concentration problems than those who don't.[112]

Not only are there too many cars in the United States and world; US manufacturers have also increased vehicle size far beyond any functional requirements and added features that bring out the worst in human behavior. They ruin quality of life for countless communities—especially lower-income and racially segregated neighborhoods. And every year in the United States, motor vehicles rank among the top ten causes of death. (Though hardly the only big offender, the 3.5-ton, aggressively designed Tesla Cybertruck has become a poster child for all these dangers. Consequently, Europe has imposed an almost total ban on Cybertrucks.)[113]

Furthermore, given their astronomical numbers, motor vehicles are among humankind's most ecologically destructive inventions. Based solely on the urgency of keeping this planet livable, any ecologically sane society must break free of automobile dominance and leave the personal vehicle in the salvage yard of history. That will be one of the most important accomplishments of degrowth, and it will improve the lives of countless people.

North America's embrace of the personal car—followed by its adoption on other continents—set in motion a series of transformative processes, most of which made life worse for most people, whether they owned a car or not. But now that a phaseout of petroleum use is an existential necessity, those changes can and must be reversed. Urban and suburban streets can be transformed from danger zones back into what they once were: places for not only getting from point A to point B but also for socializing, working, hauling, engaging in recreation, commercing, politicking, or catching a bus or trolley. Public transportation can flourish. Land that has spent decades beneath concrete or asphalt can be liberated and green spaces restored. Urban heat islands, smog, and traffic jams can become things of the past.

AUTOMOBILE SUPREMACY

Many of the ills inflicted on US society by the private vehicle are written into our legal systems. In a 2020 article titled "Should Law Subsidize Driving?", University of Iowa law professor Gregory Shill discussed in exhaustive detail the many ways in which federal, state, and local governments impose what he calls "automobile supremacy" on US society at large. Laws that privilege the interests of cars and drivers over everything and everyone else are strictly enforced, while those meant to protect us from vehicles are often laxly implemented, if at all.[114] "Crashworthiness" regulations for motor-vehicle models are designed to protect only the driver and passengers; those outside the vehicle are on their own. Drivers routinely flout laws requiring them to yield to pedestrians and are rarely charged for doing so. The most widely apparent example of what Shill calls "bright-line rules that aren't"—that is, laws that go largely unenforced—are those that specify speed limits. The Federal Highway Administration finds that typically, in the absence of traffic jams, 50 percent of vehicles are moving faster than the posted speed limit. This lack of enforcement has deadly consequences. The US National Transportation Safety Board counted more than 360,000 speeding-related fatalities from 2005 to 2014—an average of almost one hundred people per day.[115]

At the same time, other laws are strictly enforced, including those that require businesses to provide extravagant numbers of free parking spaces, snatching land away from communities and nature. Likewise with zoning laws that create urban sprawl and a need for even more infrastructure catering to the automobile. Automobile supremacy, writes Shill, is on vivid display when we observe disparities in the enforcement of laws applying to vehicle drivers versus users of public transit. While most people who are caught boarding subways or light rail without paying are arrested, lawbreaking car drivers usually receive only a traffic ticket. And, Shill adds,

> Ironically, delaying fifty bus passengers by temporarily parking [one's car] in the bus lane might be punishable by a $50 fine, but boarding that same bus with an expired pass can trigger jail time. Motorists also enjoy more constitutional privacy protection in their car trunks than people riding the subway with respect to their belongings.[116]

Most horrifically, a 2015 study found that hit-and-run drivers who killed their victims were being charged with homicide only 7 percent of the time.[117] Pedestrian crashes are consistently among the top ten causes of death for US residents under age 55. In 2021, the 7,388 pedestrians killed by motor vehicles in the US approached the combined number killed (8,005) across all 37 other countries from which the UN collected data.[118]

Your personal risk of being killed by a vehicle when you're on foot depends, in part, on who you are. The 20 percent of US census tracts with the lowest incomes have a pedestrian death rate three times as high as that in the 20 percent highest-income tracts. And from 2018 to 2022, Black pedestrians were killed by cars at more than double the rate of white pedestrians. For Native people, the rate was more than triple that for whites.[119]

Victim blaming multiplies the injustice of pedestrian deaths. When a motor vehicle crashes into a person, it's most often the pedestrian or wheelchair user, not the car driver, who's blamed. Victims are said to be at fault because they were "jaywalking" or wearing dark clothes at night or just walking, legally, alongside busy, dangerous thoroughfares—streets that aggressive drivers and lax law enforcement rendered hazardous in the first place. Describing the origins of the term *jaywalking*, Shill writes that in the 1920s, "A 'jay' was a hayseed . . . who did not know how to walk in a city; the closest epithetic analogy today might be 'hick' or 'redneck,' with all the elitism and classism embedded in those terms."[120]

A century later, this form of street crossing is still widely practiced despite being stigmatized and criminalized. But in 2024, New Yorkers, world-renowned for their jaywalking expertise, caught a big break when their city council decriminalized the practice. The decision was prompted by stark evidence of racially discriminatory enforcement of jaywalking laws: In 2023 and early 2024, 93 percent of pedestrians who received citations were Black or Latino.[121]

In most US jurisdictions, when you're crossing a street at an intersection, you're legally considered to be in a crosswalk, whether stripes are painted on the pavement or not. Drivers who fail to yield to walkers are breaking the law in most places, but they are rarely charged or fined. In a survey of pedestrian safety professionals in 171 cities across North America, 77 percent of respondents said such failure-to-yield laws were "almost never" enforced in their cities.[122]

In an interesting experiment carried out in Las Vegas in 2015, researchers recruited Black and white participants to individually and repeatedly cross two arterial streets using midblock, well-marked crosswalks at which drivers are required to yield to pedestrians—one in a high-income neighborhood and one in a low-income neighborhood. Researchers recorded the numbers of cars that did and did not pass through the crosswalk while the pedestrian was in the same half of the street (in effect, they were observing whether or not the driver potentially violated Nevada's failure-to yield law). In the high-income neighborhood, drivers were *seven times* more likely to drive into a crosswalk occupied by a Black person than by a white person. There was no such difference in the low-income neighborhood.[123]

The bulk of pedestrian deaths from car crashes occur on long, straight stretches of urban or suburban multilane roadways with cramped (or no) sidewalks, long distances between safe crossing points, and too-high speed limits, laxly enforced.[124] In other words, people on foot are dying at extraordinarily high rates because urban-transportation systems are single-mindedly designed to move the maximum number of personal vehicles through densely populated areas as quickly as possible.

The overwhelming majority of car-pedestrian crashes are unintentional; however, a growing number of drivers are deliberately threatening the safety of pedestrians, cyclists, and other drivers. Road rage, an especially dark manifestation of the general self-aggrandizement that car culture encourages, has become increasingly common in the United States. Numbers of deaths and injuries from shootings connected to road rage encounters began rising in 2018, accelerated as the Covid-19 pandemic roiled US society, and in 2025 stood at approximately double their 2018 level.[125]

A US National Highway Traffic Safety Administration report locates the roots of road rage in the driver's anonymity and sense of invincibility. "A motor vehicle," the report says, "insulates the driver from the world while, at the same time, traveling through it." Consequently, "When emboldened by the seemingly invincible power of a motor vehicle, a driver's feeling of anonymity can result in extreme rudeness and even transform an otherwise nice person into a dangerous, raging individual."[126] That makes sense, but the recent spike in such violence can't be blamed entirely on automotive supremacy. Although road rage has had a long, largely nonpartisan history, it became far more partisan in the 2020s, as US society cracked apart over public health policy, police violence, and the white-right threat to peace and justice.

With the police murder of George Floyd in summer 2020 and the subsequent waves of marches and demonstrations that swept the nation, angry motorists began attacking Black Lives Matter protesters in unprecedented numbers. During a sixteen-month period from 2020 to 2021, drivers deliberately rammed into groups of protesters a whopping 139 times, according to an analysis in *The Boston Globe*. Three people were killed and at least one hundred were injured.[127]

Crashing vehicles into crowds had been employed as a terror tactic by Islamic State adherents and sympathizers in the early 2010s, primarily in Europe. Scattered ramming attacks on Black Lives Matter protests, incited in part by postings on right-wing websites and social media, began occurring in the United States in 2015. The most notorious incident of this period came on the day of the infamous 2017 Unite the Right rally in Charlottesville, Virginia, when a neo-Nazi named James Fields drove his car into a large nonviolent group of counter-protesters, killing one person, Heather Heyer, and injuring thirty-five people.

Following the 2020–21 wave of attacks, the number of incidents has declined, but they haven't stopped, and little is being done to prevent them. On the contrary, state legislatures in Iowa, Oklahoma, and Florida have passed laws that exonerate drivers who plow into protests if they claim to have been frightened for their own safety at the time. Unfortunately, criminal-justice systems were already failing to hold crowd rammers accountable even before those driver-amnesty laws were passed. Of the 139 vehicular assaults on protesters documented by the *Globe*, drivers were criminally charged for only 65 of the assaults, and in only 4 was the driver convicted of a felony.[128] During the June 2025 wave of protests, Florida Governor Ron DeSantis made clear that vehicular assault on protesters, whatever the circumstances, was legal in his state: "If you drive off and you hit one of these people, that's their fault."[129]

THREE HUNDRED SQUARE FEET OF ASPHALT
FOR TWO TONS OF METAL

Dramatically reducing the number of vehicles on the streets will drive down the pedestrian death toll. And a degrowing society that has shaken off automobile supremacy will be better off in many other ways as well. For instance, the personal vehicle will no longer dictate that millions of acres of urban and suburban landscape be entombed under concrete or asphalt solely for the storage of cars that are not in use at the moment.

Gregory Shill, the author of "Should Law Subsidize Driving?", notes that the historical meaning of the word *parking* referred to the planting of trees, flowers, or grass in a particular space, when establishing parkland; indeed, that sort of work was done by municipal "parking agencies." Later, transportation planners adopted the word *parking* as a euphemism, Shill writes, for places to store cars, "given the positive connotations of a park (clean, natural, beautiful, etc.)."[130]

The park-themed branding was doomed to fail, and parking spaces quickly became a universal sore spot. On the one hand, drivers hate them because there's never enough; typically, one-third of cars moving through downtown traffic are just seeking a parking spot.[131] On the other hand, quality of life advocates hate parking spaces because there are way too many of them.

At any one moment, the United States is littered with tens of millions of cars that aren't being used. Businesses and local governments, in their rush to accommodate those idle vehicles, are undercutting the fulfillment of many societal needs as they pave over land that could have much higher uses. Housebuilders, too, get in on the act; in most years, they construct more three-car garages than one-bedroom apartments.[132] Henry Grabar, the author of *Paved Paradise: How Parking Explains the World*, observes,

> There are likely between 1 and 2 billion parking spaces in this country, enough to pave a small state . . . A study of 27 mixed-use US neighborhoods concluded that parking was, at peak times, oversupplied by 65 percent. Among neighborhoods with self-proclaimed "parking shortages," the oversupply was *still* 45 percent. So if there's all this parking, why is it so hard to find a damn spot? . . . Put another way, how do we break the cycle in which a perceived parking shortage requires us to deform our architecture, demolish our historic neighborhoods, and surrender our public space to satisfy the parking gods?[133]

The problem of parking is self-reinforcing. Suppose that to accommodate a parking shortage, a shopping district adds additional spaces, covering three hundred square feet of valuable land with asphalt or concrete for each space.[134] In the end, writes Grabar, the problem is still not solved:

> More parking encourages more driving, by incentivizing car ownership, pushing locations farther apart, and impairing the creation of safe, efficient infrastructure for transit, bikes,

and pedestrians. So, adding parking supply doesn't necessarily make it easier to park, especially when that parking
remains free, divided between uses, and hard to find. Until
you build so much parking that there's no longer anything
worth driving to.[135]

Local governments create more problems by requiring businesses and other properties to provide minimum numbers of
off-street, free parking spaces, based on the enterprise's type and
size. It doesn't matter if half or more of the spaces stand empty
most of the time; that's the law. Under such mandates, many businesses occupy more parking space than floor space. As a result,
Shill writes, "Parking 'craters' form that hollow out neighborhoods and destroy urban vitality."[136] It makes for a decidedly bleak
environment, one that's especially immiserating for anyone who
wants to get around on foot or a bike. It pushes people to move
even further out from the urban center, creating even larger tracts
of low-density sprawl and even more profound car dependence.

This aspect of automobile supremacy has had all-too-pre-
dictable results. If all brought together in one place, the United
States' total "impervious surface area" (which includes buildings
and other structures in addition to roads, parking lots, and other
pavement) could fully cover the state of Ohio. That's about twice
as much hard surface per capita as France or Italy, two and a half
times as much as Japan, and three times as much as Germany,
home of the Autobahn.[137]

Within US cities, 30 percent of the surface area is covered just
by streets and parking lots.[138] In summer, the sun heats up all that
exposed asphalt and concrete, which gradually releases its heat into
the environment at night. This exacerbates the "urban-heat island
effect," raising night temperatures several degrees above those in
nearby rural areas. In a future that's free of most personal vehicles,
much of that pavement will be torn up and the land converted to

green space with lots of trees, which dramatically reduce urban summer temperatures. Such conversion provides the greatest benefits to low-income urban neighborhoods, where residents are typically surrounded by much more heat-trapping pavement than people in wealthier, leafier neighborhoods. Low-income communities can be five to twelve degrees hotter on summer nights than other residential areas in the same municipality; consequently, their residents suffer more health problems and mortality during heat waves. They are also more vulnerable to flooding, because with so much impermeable surface, excess water can't percolate into the ground.[139]

The essential prelude to the de-paving of cities, clearly, will be to stop laying new concrete and asphalt. But with existing roads and highways coast-to-coast in dire need of repair, Congress continues to provide lavish funding for constructing even more roadways. Of the $350 billion earmarked for highways in the 2021 Bipartisan Infrastructure Law, about one-fourth of the funds spent as of mid-2024 had gone to highway expansion or new highway construction rather than maintenance or repair of existing roads.

On the brighter side, both the infrastructure law and the 2022 Inflation Reduction Act directed funding toward tearing up and removing stretches of urban highway long regarded as cruel monuments to racial injustice. In the 1950s and 1960s, municipal leaders, whose explicitly racist zoning laws had been overruled by courts, came to view the planning and construction of the US Interstate Highway System as a golden opportunity to cement residential segregation in place. In city after city, planners drew maps that would route multilane, controlled-access interstates on long-established dividing lines between Black and white parts of town. In other cases, the expressways were plowed straight down the middle of thriving Black communities to disrupt their development and clear space for gentrification. Residents who stayed put in neighborhoods flanking the new roads saw their quality of life severely degraded by noise and air pollution. Today, for people living within one mile of a highway, the risk of respiratory illness is almost three and a half times higher than for those living more than ten miles from it.[140]

Describing one of the most infamous cases of community destruction through roadbuilding, Deborah Archer writes that construction of a gigantic I-95 interchange in South Florida in the 1960s

> tore through the center of Overtown, a large and vibrant Black community considered to be the center of economic and cultural life for Black people living in Miami. A single massive interchange took up forty square blocks, devoured the Black business district, and took the homes of about ten thousand people. The destruction of Overtown was the realization of a decadeslong campaign by white business leaders to remove Black residents and claim that land to expand Miami's central business district.[141]

In 1950s Atlanta, the planning bureau mapped Interstate 20 through the city's West Side, creating a barrier between Black neighborhoods to the north and white ones to the south. The bureau later came clean, writing that "there was an 'understanding' that the proposed route . . . would be the boundary between the white and Negro communities," purportedly to protect the former from the latter. In Pittsburgh, Interstate 579 cut a Black neighborhood called the Hill District off from the adjacent downtown area; over the succeeding decades, the district lost four-fifths of its population and four hundred local businesses. Similarly, Interstate 4 sealed the Black West Side of Orlando off from the central business district.[142] And the legendary urban planner Robert Moses, writes Archer, "took great pains to build New York's roads and highways in a way that would limit the ability of poor people of color to visit the parks and beaches he built."

Routing of urban highways perpetuated racial segregation, just as white politicians of the era had intended. A study of 133 US cities between 1970 and 1990, at a time when the Interstate Highway System was maturing, found that predominantly Black census tracts located closer to interstates saw larger net movement of white people out of the tract and greater numbers of Black people moving in.[143] The targeting of Black and Latino neighborhoods with highway projects continued long after completion of the interstate system. Decades later, some of these roads have finally been targeted for removal, but the sums Congress allocated will be far from sufficient to get rid of many of them.[144] Meanwhile, more new miles and lanes of highway are being built.

In 2023, *The Washington Post* reported that a coalition of almost two hundred groups was pushing for a nationwide moratorium on expanding highways—citing their environmental harm and the forced relocation of nearby low-income communities of color.[145] But the movement to demolish racially unjust roadways and defund highway expansion came under attack in early 2025 when the Trump administration ordered federal departments and agencies to stop considering environmental justice when developing policies and programs.[146]

BULKED UP AND BATTLE READY

The average weight of US vehicles has been steadily increasing since the late 1990s. This has occurred primarily because of the personal-vehicle market's shift away from sedans toward larger pickup trucks and SUVs. For automakers and dealers, the bigger the vehicle, the greater the profit margin. In earlier times, therefore, salespeople earned higher commissions by pushing big, luxurious, gas-guzzling sedans and town cars. But since the 1990s, federal tax and regulatory laws have provided strong incentives for both selling and buying gas- and diesel-guzzling "light trucks" (the now anachronistic but still official term for the category of vehicles that includes all pickups and SUVs). And the bigger and heavier the vehicle, the stronger the incentive. The takeover has been sweeping; these increasingly bulked-up vehicles now account for more than 80 percent of all new vehicle sales.[147]

"Light" trucks have been supersized to the point that many are having trouble fitting into standard parking spaces and house-builders are adding square footage to garages. Owners and drivers of auto haulers—those long, double-decker truck trailers that carry up to nine vehicles at once—are having trouble staying under the federally mandated eighty-thousand-pound weight limit. They're increasingly forced to carry fewer vehicles per load, which costs them dearly. The industry is lobbying for an eight-thousand-pound increase in the weight limit, but experimental studies and the laws of physics say that raising the weight limit by just that 10 percent would increase damage auto haulers inflict on road pavement (which, as one might expect, is already quite severe) by almost 50 percent.[148] With such an increase, Congress would need to allocate even more money toward highway repair.

Since around 2010, automakers have been turning out pickups and SUVs that are not only larger but also project an increasingly menacing, even militarized, image. Towering front ends armored with "bull bars" on the front end, blackened windows, jacked-up suspension systems, giant tires, and often loud exhaust systems, are intended to convey a road-ruling mystique. In one ad for the Ford F-150 pickup, the term "military-grade, aluminum-alloy" appeared five times. In 2021, a writer for *Vice* pointed out that some leading pickup models had become almost as large as World War II–era tanks.[149]

Angie Schmitt, the author of *Right-of-Way: Race, Class, and the*

Silent Epidemic of Pedestrian Deaths in America—a heartbreaking account of the carnage being wreaked on our roadways—has written elsewhere about some of the reasoning behind these trends: "Personal vehicles are not merely functional appliances: They are used as refuges, fortresses and private enclaves, and serve as important signifiers of class and gender identity."[150] That has trapped customers in a feedback loop of self-preservation, an arms race if you will, that has supercharged the sales of bigger and bigger and meaner and meaner pickups and SUVs. Auto dealers and their salespeople are offering the kind of height, heft, and armoring that will maximize the safety of the customer sitting across the desk from them while, as a side effect, threatening the safety of other drivers—some of whom will soon be visiting that same showroom, shopping for tank-sized vehicles to protect their own families.

Even as they discomfit drivers, riders, and pedestrians in their vicinity, today's pickups and SUVs are all about maximum comfort for their own drivers and passengers. Automakers have steadily sacrificed cargo space in a competition to offer customers the industry's most luxurious "living rooms on wheels."[151] A 2016 critique of this trend in the publication *n+1* noted that at that time,

> Fully 90 percent of full-size trucks are ordered with seating for five and two-thirds with four doors. And as the cabs get bigger, the beds get smaller: these enormous trucks are too small for 4×8-foot sheets of plywood, but they offer a wide array of extravagant options, including electric tail gates operated from the key fob and a step ladder to get into the back, even though everyone knows that the proper way to mount a truck bed is to step up onto the tire.[152]

As trucks became more suburbified, truck makers' advertising went the other way, increasingly populated by rugged individualists taking the road (and the off-road) less traveled. The tough guys featured in the ads—advertisers' notions of "real Americans," including ranchers, oil field roughnecks, hunters, firefighters, farmers, loggers, and guys who just like to drive to the tops of rock piles—today account for a small minority of the truck-owning population. Rather than towing cattle trailers out on the range, most pickups are cruising along city streets and interstate highways, sporting empty beds and spotless tires, their drivers settled into cushy captain's seats.

ASSAULT AND BATTERIES

The need to drive vehicle supremacy out of US society has become an even more urgent matter of life and death for anyone who's not buckled up inside a big, well-armored truck. In addition to being heavier by far than passenger vehicles of the past, the new breed of full-size pickups and SUVs have extra-stiff frames, towering front ends, and, often, a raised-suspension system. Those features combine to pose a much greater danger to pedestrians. A bigger truck slams into a person with greater force given its excessive mass. It also strikes higher on an adult's body than a sedan would—around mid-torso and maybe the head—thereby causing the greatest damage where vital organs are located. Furthermore, a person struck by a sedan is usually thrown up onto the hood, whereas the tall front end of a truck causes much worse injury by slamming into the upper body. This often causes the victim to fall backward to the pavement and often be swept underneath the vehicle—a worst-case scenario.[153] The results have been all too predictable. While overall pedestrian fatalities shot up by 71 percent between 2010 and 2023, the subset of fatalities caused by pickups and SUVs climbed precisely twice as fast, at rate of 142 percent.[154]

Walkers, cyclists, and even other drivers are at higher risk than ever. In a head-on collision with a sedan, big trucks—especially those with raised suspension and oversize tires—can crawl up onto the car's hood and smash through the windshield. A study of head-on crashes involving 838 vehicles found that when a pickup or SUV collided with a sedan, the odds of death were a shocking 7.6 times higher for the sedan driver than for the truck driver. And those data were collected between 1995 and 2010, before the continued surge in truck weight and height made them even more lethal.[155]

Degrowth requires a deep reduction in the total number of motor vehicles, not just a switch to vehicles with lower greenhouse gas emissions. Swapping out America's 278 million gas and diesel passenger vehicles for a similar number and range of electric models wouldn't solve any of these problems, and it would exacerbate some of them. A large majority of the EVs now rolling off assembly lines are big pickups and SUVs, not small, energy-efficient—and always less profitable—electric sedans. Thanks to their batteries, electric trucks are also much heavier than their internal-combustion cousins. The battery in Ford's flagship F-150 Lightning pickup weighs 1,800 pounds and is the size of two mattresses. That pushes the Lightning's weight up to an astonishing 6,500 pounds, making it 35 percent heavier than the gas-powered F-150. To be fair, the Lightning does weigh 28 percent less than the nine-thousand-pound Hummer EV, whose battery alone outweighs an entire gas-powered Honda Civic.[156]

Battery-powered pickup and SUV models pose an even greater threat to life and limb than their fuel-powered siblings, and that's not just because of their greater weight. They also achieve much faster acceleration than fueled models, thanks to their efficient motors. For example, the Lightning pickup goes from zero to sixty miles per hour in a blistering 3.5 seconds, while one of Tesla's SUVs does the same in 2.5 seconds.[157] That kind of acceleration, research shows, is associated with increased risk of a pedestrian crash.[158] On a city street, a quiet electric truck capable of that sort of sprint can seem to come out of nowhere. Furthermore, with the combination of greater weight and potentially higher speed, a large EV driven recklessly could cause especially severe damage to the human body. Each doubling of speed quadruples the force with which the vehicle strikes the body.

Electric vehicles pose most of the same societal problems that internal-combustion vehicles do. They too require vast acreages of concrete and asphalt. There's no reason to expect commuting in an EV to be any less irksome than commuting in any other vehicle. And EVs are linked to all the same injustices. They have the same suppressive effect on public transportation. They too prohibit development of lively urban street life and feed urban sprawl. Their drivers are no less susceptible to road rage. An EV could pose an even greater threat to pedestrians and cyclists than an internal-combustion vehicle of similar size. And the EV's one claim to superiority—lower environmental impact—is not all it seems to be.

To refer to EVs as "zero-emissions vehicles" is an error. True, they don't have tailpipe emissions or even a tailpipe. But they cause the release of greenhouse gases into the atmosphere in many other ways. Almost all EVs in today's global fleet draw on an electricity supply that's generated in most cases by coal- or gas-fired power plants or by other nonrenewable sources, so those carbon and methane emissions and their other environmental impacts must be charged against the EV.[159]

To arrive at a fair comparison between EVs and other vehicle types regarding overall environmental impacts requires doing what industrial ecologists call a life-cycle analysis. There is a growing body of research generating such analyses, accounting not only for energy supply to the vehicle and resultant greenhouse gas emission but also for the ecological impact of much else that's associated with the two vehicle types: the manufacture and the extraction of resources that it requires; operation throughout their working life; and end-of-life processing and disposal. A review paper published in 2023 examined fifteen such studies, comparing EVs' lifetime climate impact with that of gasoline- and diesel-powered vehicles. EVs' impact was lower, but very far from zero. Their lifetime greenhouse gas emissions were about half of those from corresponding gas vehicles and almost 60 percent of those from diesel vehicles.[160]

When it comes to ecological effects that go beyond climate disruption, EVs tend to score even more poorly than their fueled predecessors. The biggest source of problems is that big battery, the manufacture of which has touched off a global race to mine and process enormous quantities of lithium, cobalt, and other metals required for production of batteries, light enough for use in powering cars and trucks. All steps—mining, processing, and battery manufacturing—cause ecological destruction; pollute surrounding communities' air, water, and soil; entail the exploitation and endangerment of workers; spark local violence; and run the risk of "green-resource wars" that could be touched off by competition among the world's military powers—mostly in low-income countries. Considering climate disruption in conjunction with other environmental issues, the message is clear: For every motor vehicle not manufactured, no matter whether it would have been fuel- or electric-powered, prospects for a better quality of life will improve in the Global South as well as the North.[161]

There's yet another, much less obvious way in which a takeover of the automotive market by EVs would exacerbate the ecological damage that's done by that market. David Zipper, who in recent years has been documenting the downsides of personal vehicles in general and EVs in particular for *Slate*, *The Atlantic*, *Vox*, and other outlets, raised alarms in 2023 over the tire pollution they cause. He was pointing not to the hazards created by disposal of worn-out tires (which *is* a huge problem) but to the product of the wearing-out process: tiny particles of rubber, nylon, and other materials that are scoured off tires by the road surface. The global quantity of this diminutive detritus adds up to as much as six million metric tons annually. Zipper cites a study estimating that tires produce as much as 28 percent of all microplastics going into the oceans, where they work their toxic way up the food chain. Regarding impacts on humans, he writes, "Various tire components have been linked to chronic conditions including respiratory problems, kidney damage, neurological damage, and birth defects—a particular concern in neighborhoods adjacent to highways, whose residents skew low-income and minority."[162]

And it turns out that electric vehicles produce more—maybe a lot more—particulate pollution than other vehicles, because their tires wear out faster. Indeed, rapid tread wear is EV owners' number one tire complaint; an auto-supply analyst told Zipper, "They're expecting forty thousand miles out of their tires, and they're getting thirteen thousand." Why are EV tires losing their tread so fast? Part of the reason involves yet another conflict between energy efficiency and ecological impact, one that echoes the story of LED lighting. EVs are programmed, as an efficiency enhancement, to slightly engage their brakes while in motion as a means of recycling energy to the motor. That increases friction between tire and pavement, rubbing off more of the tiny particles. Other factors are the great heft and rapid acceleration of large EVs, which also increase friction.

IT'S EITHER US OR THE CARS

Electric vehicles exemplify how attempts to address a highly complex, deeply rooted problem by fixing just one of its elements can backfire elsewhere. Just as the switch to energy-efficient LED lighting ended up making light pollution worse, a wholesale conversion to lower-emission EVs could put pedestrians in even greater peril and invite environmental and human rights abuses around the world. Potential solutions have been proposed for all the problems that car culture leaves in its path, but, whether they're deployed individually or collectively, they can't halt or repair the overall damage wreaked by automobile supremacy.[163]

For instance, streetscaping for pedestrian safety is clearly needed, but doing it in one area doesn't reduce overall motor vehicle use. Traffic congestion is simply displaced to other, often less privileged, areas. David Owen makes this point more broadly in his 2011 book *The Conundrum: How Scientific Innovation, Increased Efficiency, and Good Intentions Can Make Our Energy and Climate Problems Worse* with this aphorism: "Traffic congestion is not an environmental problem; traffic is." In a car-saturated society, he argues, efforts to relieve congestion—adding and widening roadways or adding more bus and rail to the transportation system—provide only fleeting relief.

Dense traffic, Owen argues, is perhaps the only thing deterring humans from driving even more than we do now. Smoother traffic flow makes driving more attractive, or at least more tolerable, and lures more cars back onto the roadways. Owen has a pithy proverb for that, too: "Road space begets road use." Public transit, in contrast, can reduce personal vehicle use, but, he adds, only if it's accompanied by a reduction in the amount of road space allotted to cars. That can be accomplished by turning entire lanes or streets over to bikes, buses, and rail.[164]

There's no path to ecological health and human flourishing that doesn't include freeing society from automobile supremacy. That liberation will require sweeping changes in transportation and the built environment, but the improvements in quality of life will be profound and lasting.

"CALCULATED MISERY" AT THIRTY THOUSAND FEET

As automobile supremacy withers, degrowing societies will become better places to live for almost everyone. Air travel, another mode of transportation targeted by climate activists, is probably even more notorious than road travel for providing passengers with an end-to-end miserable experience. It's also ruinous to the quality of life for people living anywhere in the vicinity of airports, especially under takeoff and landing paths. A deep reduction in air travel, essential to preventing climate catastrophe, will also accelerate our collective evolution toward a better life with degrowth.

A phaseout of fossil fuels cannot be accomplished without deep reductions in air travel. Ideas for keeping aircrafts aloft long into the future—by using hydrogen or biofuels, or impossibly lightweight batteries or even mini-nukes—will remain in the realm of fantasy. With a phaseout, the reduced quantities of oil and gas that the world will continue to need and use will have to be directed largely toward food and fertilizer production and the manufacturing of essential materials. That will leave little or no fuel available for hurtling people through the atmosphere at six hundred miles per hour or dropping bombs on each other. But I offer this prediction: Most of us will adapt quickly to living our lives on the earth's surface. And those who follow us will look up at the starry sky from time to time and think, "I can't believe they put up with all that nonsense just to get somewhere fast."

Andre Gorz's observation about private vehicles—that, like villas on the Riviera, they improve their owners' quality of life only if they belong exclusively to a small, elite class—fits air travel just as well. The many miseries of flying are all too familiar to those who've had the experience, and air travel has provided ample fodder for stand-up comedy, at least since the early 1990s when flying became more broadly accessible to the middle class.[165]

The list of happy subtractions that will come with the phaseout of air travel is extensive. It includes long lines and coercive herding at check-in, security, and boarding; flight delays and cancellations; crowding and discomfort on board; baggage hassles; and much more. Then there's the cast of characters who make one's flight more memorable: the deep-reclining guy in the seat ahead; the seat-kicking kid in the seat behind; the loudmouth across the aisle; the guy stealing liquor off the cart and indulging in vociferous self-expression.[166]

Two years before the September 11 terrorist attacks, in a presentation to the American Psychological Society, Irwin Sarason and Jonathan Bricker of the University of Washington unveiled the results of what they said was the first-ever study of stress related to air travel. They uncovered plenty of it among frequent economy-class business travelers and received months of media attention.[167] In 2002, Sarason and Bricker were back with another study. This time they found that the percentage of passengers who said they were "somewhat, moderately or severely stressed" by commercial airplane travel—a figure that had stood at a strikingly high 60 percent among those who flew before 9/11—shot up to 81 percent among those surveyed a few months after the attacks. Interestingly, the most severe stresses were prompted not by fear of terrorism but by the airport-security procedures introduced in response to 9/11.[168]

In the years since the internet began empowering travelers to shop around for the lowest fares, airlines have been extracting more money from passengers by charging for sitting together with loved ones, checked luggage and even carry-on bags, and other services that used to be included in the ticket price, by offering "upgrades" for earlier boarding or better seating. But with most passengers saying "no thanks" and sticking to the basic service, the companies realized they had to increase the incentives to upgrade by making the basic service even more unpleasant and annoying. So now they're imposing what Columbia law professor Tim Wu has dubbed "calculated misery"[169] on the passengers in the rear of the plane, as a way of prodding them to pay extra fees just to get back the three more inches of legroom that they used to get for free or to sit next to their family members as they once did without having to pay for the privilege.[170]

Economy-class air passengers can be seen as victims of predatory corporations that cater only to the wealthy fliers up front who keep the profit stream flowing. But shed no tears for those sitting in the sardine-can section. Although 90 percent of Americans have flown at least once in their lives, frequent flying is a luxury that only a small minority of the world's people can afford. To engage in routine air travel is to belong to a global elite, whatever your boarding-group number. An estimated 12 percent of Americans take two-thirds of all domestic flights. Global disparities are even more stark. Fewer than 20 percent of the people in the world have ever been on an aircraft, and just 5 to 10 percent fly any given year, according to the climate group Possible.[171]

The hassles of flying have been exacerbated by a sharp increase in air-rage incidents. Air rage runs through a spectrum that includes refusal to comply with safety instructions; abuse aimed at cabin crew or other passengers; assault; physical damage to the aircraft; and damage to airline or employees' property.[172] Global numbers of air-rage incident reports increased steadily from 2007 to 2019. Then came the Covid-19 pandemic. In 2021, with high-volume air travel having resumed and right-wing outrage over Covid regulations peaking, the Federal Aviation Administration (FAA) received nearly six thousand "unruly passenger" reports, far more than in previous years. The rising tide of air rage featured people hitting or biting flight attendants, attempting to stab fellow passengers, breaking someone's teeth, and sliding unbidden down the evacuation chute.[173]

An analysis of 228 air-rage incidents occurring between 2000 and 2019 found that three-fourths of the passenger-perpetrators were male. More than half of the incidents were categorized as "physically disruptive" or "life-threatening" behavior. Among fifty-six incidents for which details were available, ten involved sexual harassment, eight involved personal space issues, six were physically violent, and three were terrorism related. Others involved attempts to open an aircraft door, to damage aircraft equipment, and to dispute seating arrangements. There was also wearing of a red MAGA hat, homophobic slurs, and, predictably, anger over a fellow passenger's flatulence.[174]

The miasma of bitterness that pervades air travel is deepened by the exemptions that upper-class and business travelers enjoy. The journalist Maia Szalavitz has pointed out the role that wealth and class play in airborne immiseration, most obviously during "the long walk through the cushy first-class cabin to a claustrophobic middle seat at the back of 'torture class.'" But, she adds, the indignation runs both ways:

> In fact, a study published last year suggested that economy passengers feel the most rage when they walk through first class: it reminds them of their diminution. But first-class passengers weren't any calmer: those subjected to the indignity of having the unwashed walk through their space, rather than boarding out of their sight through a middle door, were even angrier.[175]

But enough with the air travel already. As we head toward degrowth, may we see both air rage and road rage recede swiftly in our rearview mirror.

ON THE RIGHT TRACK?

Rail is widely considered to be an environmentally superior substitute for intracontinental air travel on the grounds that it uses less energy per passenger, per mile traveled. But for many obvious reasons, rail can't support the volume of long-distance travel that the airline industry currently achieves. Furthermore, David Owen contends, whatever the scale of rail's advantages over air travel in energy use and climate impact (and even those may be much smaller than advertised),[176] "the main environmental benefit of train travel is that it takes longer, making people less likely to do it."[177]

There has long been a hope among green thinkers that the widespread deployment of high-speed rail lines would lure passengers away from air travel. Others argue that a society in the process of actively phasing out air travel should opt not for high-speed rail but for good old snail rail. By making long-distance travel more attractive, the reasoning goes, super-fast trains would only encourage people to travel more. Owen cites an example from his own experience with high-speed rail:

> The Eurostar [rail line] made it easy for me, when I was in London, to hop over to Paris for lunch and the latest exhibit at the Louvre, ho-hum. The environment would have been better off if, instead of doing that, I had stayed in London and walked from my hotel to the British Museum. And it would have been better off still if I'd skipped London and stayed at home, on my side of the Atlantic.[178]

High-speed rail infrastructure requires much bulkier, more elaborate infrastructure, requires far more material resources, especially concrete, and costs far more per mile to construct than traditional electrified rail. Efficient versions of conventional electric rail can reach many more places for the same cost and lower environmental impact.[179]

HAUL LESS FREIGHT, MORE PEOPLE

Disparate elements of the growth economy interact in count-less combinations to inflict harm, sometimes in unexpected ways. Reducing the quantities of energy and material resources consumed by one type of economic activity can boost another, seemingly unrelated, activity that reduces resource consumption and may also spin off some beautiful surprises.

For example, reducing superfluous production of goods, thereby lightening the economy's material footprint, as discussed in chapter I, will trigger all sorts of salutary ripple effects. It could even have an unexpected side effect: helping end automobile supremacy by fostering an increase in rail travel.

Today, the National Railroad Passenger Corporation, known to us as Amtrak, is stuck in a Catch-22. It carries only a small share of the nation's travelers, partly because of its limited geographical coverage and reputation for often running late. But efforts to extend Amtrak's reach and improve service are being impeded by freight companies such as Union Pacific and Burlington Northern that own the tracks used by Amtrak (except in the northeastern rail corridor). These companies view passenger travel as an unnecessary, unpopular service that intrudes on their essential business, freight hauling, so they are undermining Amtrak's ability to expand its coverage nationwide and provide on-time service.[180]

In defiance of federal regulations, the rail companies compel Amtrak's trains to yield right-of-way to freight trains in any stretch where the two are sharing a single track (which means most of the network). As regular US rail passengers know all too well, the resulting delays can drag on for twenty minutes to several hours. Furthermore, the freight haulers keep increasing numbers of cars per train, to the point that many stretch out as far as two miles. A much smaller, faster Amtrak train is always diverted onto a siding until an oncoming freight train finally arrives and lumbers past.

It can also get stuck behind a slow, two-mile-long train for many miles with no way to pass it, because such long freight trains can't fit into a typical siding. According to Amtrak, "Freight train interference—a dispatching decision made by a freight railroad to delay Amtrak passengers so that freight trains can operate first—caused nine hundred thousand minutes (over 1.5 years) of delay in 2023."[181]

Amtrak received a big funding boost from the Inflation Reduction Act in 2022 and developed a $75 billion plan to greatly extend its reach in thirty-nine areas of the country. But the freight companies, reluctant to further share their tracks, have fought that plan tooth and rail. Expansion remains only a dream.[182]

Now imagine a US economy of the future with much lower material consumption. There would be far less freight to ship and therefore much less track congestion. According to the US Federal Railway Administration, almost half of rail freight consists of consumer goods and "other miscellaneous products," quantities of which can be deeply reduced with degrowth. The other half is made up of bulk commodities. Much of that freight, including food grains like wheat, would still need to be hauled by rail, while transport of other bulk commodities, including coal for power plants and feed grains for factory farming of livestock (see chapter IV) could and should be phased out. The economy would then have much smaller requirements for freight-hauling, and, with fewer people driving and flying, a greater need for long-distance passenger rail. The tracks can be nationalized, passenger rail can receive top priority for scheduling and track space, and thousands of miles of new track can be extended into underserved regions.[183]

Amtrak's vicious circle—getting too little funding because it draws too few passengers, while drawing too few passengers because it's underfunded and has been hobbled by the freight companies—can be broken. Passenger trains will then run mostly on time and far more frequently, leading to greater passenger satisfaction. Demand for rail travel will rise, so more trains will run, leaving fewer cars on the nation's highways and fewer planes overhead.

With degrowth, railways will be the primary long-haul transportation mode (maybe networked with sailing ships?). Green(ish) public transportation within urban areas will be expanded, not by digging under the streets but by taking over existing streets and expressway lanes, eventually displacing almost all private cars.

Transportation could be the opposite of Gorz's houses on the Riviera: systems that are useful because everyone can benefit from them while doing as little ecological harm as possible.

IV.

This Land

There can be no place in a degrowth society for a food-production system like the one that exists today in the United States. Instead, there must be a system that's adequate to produce sufficient quantities of healthful food for all while also causing minimum ecological disruption—in short, the antithesis of the current farm-food-profit economy.

The drive for maximum profit from production at maximum speed, essential to capitalism, has led to the ecological and human disaster that is US agriculture. A field sown to a food crop can't be sped up like an automobile assembly line. The entire food economy depends on the sunlight that powers plant growth, and that light arrives at Earth's surface at a rate that can't be dialed up. The ability of a farmer with a field of wheat or beans to circumvent the rhythm of the seasons or the vagaries of daily weather is strictly limited. Yet our dominant farming systems are aimed at doing just that. And the system's not working; industrial technologies aimed at juicing the output from agricultural fields in defiance of nature's limits have come at huge costs in fossil-fuel, material inputs, and environmental damage.

Agriculture's foundation is a belowground ecosystem that, when healthy, operates in equilibrium, not like a growth-addicted corporation or plucky digital start-up. Ignoring that reality, capitalism's rules require that the nation's farms feed not only human bodies but also a constellation of grow-or-die industries. Seed, fertilizer, and farm-chemical companies, fossil-fuel giants, farm-equipment manufacturers, irrigation-system installers, loan companies, real estate agencies, and a host of others, are seeking to profit on the front end. At the back end, we have an even larger cast of characters: grain handlers and traders, speculators, freight haulers, food processors, biofuel producers, packagers, food wholesalers and retailers, advertisers, and at the end of the line, two notably complementary industries: food service and weight-loss products and programs.

Without those burdens to carry on their shoulders, agricultural ecosystems and the people who farm them would be free to care for the land and nourish their fellow humans rather than feed investors' fortunes.

GOODBYE, EARTH ABUSE

Only about one in seven Americans lives in a rural area, and fewer than three in a hundred live on farms.[184] Those who make up the nonfarm, nonrural population can imagine many features of the food economy that should be left behind—mostly problems with the food itself and the way it's obtained, consumed, and disposed of. The sales pitch to urban and suburban populations for alternative, more ecologically sound forms of agriculture has therefore focused mostly on consumer appeal. The reason that you, the consumer, should want your food to be grown organically or biodynamically or in another alternative system, we are told, is because it will look and taste better, provide better nutrition, and be free of toxic chemicals.

That last claim is generally true; the first and second may or may not be the case, depending on the situation. Furthermore, giving priority to consumers' aesthetic motivations leads to a tight focus on vegetable and fruit crops, which occupy less than 5 percent of US cropland. Cereals, grain legumes, and oilseed crops, which together occupy about two-thirds of cropland, are mostly left out.

Many ecological and human-welfare issues swirl around both vegetables and grains, but the many dangers to humans and non-human nature that agriculture is creating in the places where it's practiced are not widely discussed outside of farm country. In this chapter, I'll focus on food production and the harms that can be purged from our society by moving away from annual crop monocultures, soil tillage, factory farming of animals and plants, and other harmful features of US agriculture—some of them a century old, others going back ten millennia. The greatest share of benefits from such a transformation will go to ecosystems, both below and above the soil surface, and to the people whose labor provides our sustenance. That revolution can't be achieved only by appealing to high-end consumer preferences or romanticizing food. If it is achieved, myriad food-related problems will fade into the past—for all of us.

The ecologically necessary transformation of food production, carried out justly, will bring an end to the damage now done by the industrial food chain, all the way from the cornfield to the drive-up window at Burger Hut or gourmet home-delivery services. A degrowth-oriented overhaul of the food-production system could even help heal the urban-rural political rupture that threatens to tear Western nations apart and put an end to our disastrous corporate-controlled food policy abroad.[185]

Agriculture policy is also climate policy. Food obviously can't be phased out like fossil fuels can, but excessive or wasteful food production and consumption can be eliminated as part of degrowth. And unlike fossil-energy production and use, food production can be transformed in such a way that its net effect is to remove more carbon from the atmosphere than it releases—not as well as natural ecosystems do, but, potentially, it can help.[186]

Agriculture has another advantage over the energy sector. Without deep cuts in energy demand, an extraordinarily steep scale-up of renewable-electric capacity and battery storage will be needed to compensate for the decline of energy from fossil fuels. That, in turn, will require accelerated mining, industrial development, and widespread encroachment on ecosystems, all ecologically destructive. In contrast, most efforts to mitigate agriculture's climate impact will be ecologically beneficial. They are actions we should have been taking already for other good reasons: ending soil erosion, allowing belowground ecosystems to restore themselves, curbing biodiversity loss, and stopping pollution of rivers, stream, and groundwater.

There are many less obvious ways in which reducing the harms spun off by agriculture can reduce those caused by greenhouse gases—and vice versa. For example, finding enough land area to produce ample food in urban and suburban areas could well entail breaking up and hauling away lots of pavement. That, in turn, would provide major ecological benefits while helping liberate society from automobile supremacy and reducing carbon emissions.

GOODBYE, FEED AND FUEL GRAINS

Growing and harvesting food crops whose end products are seeds or kernels—cereals like wheat and oats, legumes like dry beans, peanut, soybean, and oilseeds like sunflower—occupies far more land area than vegetable production. Food grains will therefore continue to be produced primarily in rural regions and, because they aren't perishable, they can be shipped efficiently by rail to population centers across the country.

US farmers sow approximately 275 million acres of land to annual cereals, grain legumes, and oilseeds each year. But more than two-thirds of that cropland is sown to just two species of feed grains, both consumed mostly by livestock: corn and soybeans. Much of their acreage is concentrated in the upper Midwest on some of the nation's deepest and most fertile soils. For example, the odds are better than one in three that a meteorite landing at random somewhere in the state of Iowa will be found in either a corn or a soybean field (in spring, summer, or fall) or in a field that will soon be sown to corn or soybean (if it's wintertime).[187]

Together, just four crops—corn, soybean, wheat, and cotton—occupy 90 percent of all US land sown to field crops.[188] And of those four, only wheat, which occupies 17 percent of total field-crop acreage, serves primarily as human food. The only major use of field corn for human consumption is in producing high-fructose corn syrup, a nutritionally negative food, while sweet corn is classified as a vegetable crop and isn't included in these acreages. Most of the corn and soybean harvest goes into feeding cattle in feedlots or dairy operations or to feed swine and poultry in confinement facilities. Meanwhile, one-third or more of the corn harvest is used to produce ethanol for feeding motor vehicles.[189]

If we in the United States take a sharp turn onto the degrowth path, the most direct way to lighten agriculture's environmental footprint is to deeply reduce production of those grains we can't eat. Analysts at the universities of Georgia and New Mexico have shown that for energy and climate concerns, deep cuts in fuel-ethanol and grain-fed beef production—or, even better, their elimination—would have greater positive impact than any other changes in US agriculture.[190]

If we were to produce far less corn and soybean for feeding beef and dairy cattle while choking off the supply of corn to the fuel-ethanol industry, society would gain far more than it loses. The amount of energy consumed in producing feed grains for cattle far exceeds the amount of energy contained in the beef that's eventually produced by those cattle and sold.

Unlike cattle, chickens and hogs can't live on range alone, pasture, and hay, so they need feed grains. But even those animals should be out on the land, not locked up in indoor force-feeding facilities, as most are now. In a single, typical concentrated animal-feeding operation (CAFO), eight hundred thousand swine can generate more than 1.6 million tons of waste a year—more than the annual human-waste output from the city of Philadelphia. According to the National Association of Local Boards of Health, the nation's livestock produce at least three times, and possibly as much as twenty times, as much fecal waste as America's human population. Large numbers of such facilities tend to be concentrated in areas with an appropriate climate, plenty of cheap labor, and ample access to feeds and transportation facilities. Communities unlucky enough to have all those advantages are vulnerable to having CAFOs move into their area and plunge them into air- and water-pollution nightmares.[191]

The investigative reporter Will Potter delves deep into the execrable details of the CAFO fecal waste problem in his 2025 book *Little Red Barns: Hiding the Truth, from Farm to Fable*. And it's a massive problem for that industry; Potter hears factory farmers quip that they feel as if they're in the manure business, with meat or milk being mere byproducts. The putrid, toxic slurry is pumped out of confinement buildings and hauled to industrial "lagoons" with capacities of twenty to forty-five million gallons each. Workers around the lagoons are at constant risk of death from toxic gases; inhalation can either kill them directly or cause them to lose consciousness and fall into the pit, with little chance of getting out alive.[192]

Because it's rich in nitrogen and phosphorus (and despite it being noxious and toxic), the preferred method for discarding these extraordinary quantities of lagoon slurry these days is to spray it onto fields as fertilizer. Potter writes, "Each of these sprinklers can spray up to two hundred gallons of fecal matter per minute. . . . It shot from the spouts heavy and brown, and arced into the field. I watched the brown stream turn into a brown mist as it caught in the breeze, leaving the farm for back yards, open windows, play sets." Wherever it goes, the stuff remains both toxic and infective; Potter cites research published in the *Proceedings of the National Academy of Sciences* estimating that air pollution from CAFO fecal waste kills seven thousand Americans per year, surpassing the death toll from coal-fired power plants.[193]

Factory farming produces far more manure than can be absorbed safely by surrounding farmland within economic-hauling distance. Excessive quantities of the fetid slurry end up being sprayed or spread onto fields anyway, so more nitrogen and phosphorus seep into the soil than a crop's roots can take up. The quantities that escape absorption become pollutants, percolating into ground water or running off into streams.[194]

Dangerous nitrate levels are found in drinking water in many rural areas near cropland to which large amounts of fecal waste and/or synthetic fertilizers have been applied. High nitrate levels in water increase the risk of colon and rectal cancers, thyroid disease, and birth defects. They are especially dangerous for children, robbing them of oxygen and causing "blue baby syndrome."[195]

The human toll inflicted by agribusiness begins with food production, and the exploitative food-processing industry takes over from there. Most notorious in that sector are meatpacking plants with their constant threat of physical injury. In the period 2015 to 2017, an average of two US meatpacking workers per week suffered work-related amputations. Starting in 2020, infectious disease was added to the list of grave threats. Meat processing requires workers to operate at close quarters in the very kind of chilly, dry atmosphere in which the Covid-19 virus survives and transmits most readily from person to person. In 2020, meat workers in some large plants were more than fifty times as likely as the general population to be infected by the then-novel coronavirus.[196]

Each step in the industrial food chain—on-farm production, transport, product handling, food processing and manufacturing, packaging, more transport, food retailing, and the restaurant industry—is oriented to maximize profit, whether or not workers suffer along the way or consumers obtain healthful food at the end of the chain. Generally, as plant- and animal-derived products move through the system, nutritional quality declines and ecological harm accumulates. Processing (which the US Department of Agriculture aptly refers to as "food manufacturing"[197]) and the food-handling steps that follow it—packaging, transporting, retailing, consuming, and waste handling—together account for 10 percent of global greenhouse gas emissions and 15 percent of US emissions. In today's US food system, those sectors have greater adverse climate impact than production of the food itself. They should and most certainly can be downsized, and that is eminently achievable. We got along fine with much smaller food-related industries for most of our history. Their growth is a relatively recent phenomenon, with emissions from the nonfarm sectors of the US food chain having ballooned 3.6-fold in just twenty-five years, from 1990 to 2015. A sector that grew so big so quickly can be degrown even more quickly, to the benefit of human and ecological health in this country. That will also reduce our food industries' emissions of carbon dioxide, methane, and other greenhouse gases, which, in turn, will help lessen the threat of crop failures and famine that climate disruption poses to subsistence farm families throughout the Global South.[198]

FREE UP LAND FOR PERENNIAL PLANT LIFE

When US farmers stop producing the billions of bushels of corn, sorghum, and soybeans that go into feeding cattle and biofuel facilities, a large portion of the acreage now devoted to feed grains will be available to convert into range, hay, and pasture lands for producing grass-fed meat and dairy products. Such systems, relying on perennial grasses and other perennial species grown in mixed stands, will help reverse soil degradation, which is the eternal plague of annually tilled systems, like those that produce all of today's grain crops. With such a transformation, the quantities of animal products will be much more modest than those produced today, and that will be a good thing. Americans currently consume grain-fed beef, as well as pork and poultry, at unhealthful and ecologically insupportable rates anyway. Much of the land liberated from feed grains would then be available for growing food crops: cereals, grain legumes, and oilseeds. Land would once again be respected as the treasure it is, not regarded simply as a platform for profit making.

Land abuse can take many varied forms. For example, since 2021, a confrontation in the upper Midwest involving fuel ethanol and carbon dioxide (CO_2) pipelines has pitted a grassroots alliance of landowners and land lovers against predatory industrial interests. And the struggle could have profound consequences for climate and the broader ecological emergency. The most prominent company proposing to build networks of such pipelines in Iowa, Minnesota, Nebraska, and the Dakotas is Summit Carbon Solutions, based in Ames, Iowa. If completed, the system's two thousand miles of pipe would transport supercooled, liquefied CO_2 from more than fifty ethanol-fuel plants scattered throughout the region to North Dakota, where it would be pumped into depleted oil wells. There, the companies claim, it will remain trapped indefinitely in deep rock formations. Such pipelines are crucial to a carbon-capture industry that has already been found incapable of significantly reducing the nation's net greenhouse gas emissions.[199]

Summit pressured landowners along the planned pipeline routes to sign easement agreements that would cede control over portions of their property where the pipes would run under their soil. Hundreds refused to sign, citing safety concerns, damage to their cropland and waterways, and corporate intrusion on their property. Most of the resistant landowners are farmers. Almost uniformly white, they are joined in their opposition by Native tribes. An antipipeline organizer with the Indigenous-led Great Plains Action Society told me in 2022, "It's a *very* sensitive matter for Indigenous folks. This country was founded on land stolen from them, and now they are trying to prevent some of that land from being stolen *again*, this time by big corporations. So Indigenous people are standing shoulder to shoulder with farmers."[200]

In the fall of 2023, Summit had to postpone the kickoff of pipeline construction until at least 2026, thanks to fierce resistance being put up by an atypical alliance of farmers, Indigenous people, and environmentalists.[201] In public meetings, in courtrooms, and out on the land, the grassroots were standing up to corporate power. Such are the kinds of movements that can coalesce to bring about degrowth.

LAND REFORM IS COMING TO THE USA

Lands rescued from corn, soybean, pipelines, and feedlots, whether they're repurposed for growing essential food crops or restored as prairie or forest, could be the initial focal point for much needed land reform, US-style. Decades of economic concentration, with larger and larger farms falling into fewer and fewer hands, must be reversed if our irrational food system is to be overturned. That will mean not only buying out and breaking up larger landhold-ings to create more, smaller ones. It will also bring a renewal of more diverse, vital rural cultures and economies that fit into a more ecologically well-adjusted future. With land reform, there will be an opportunity to make reparations to three groups of people: those from whose ancestors European immigrants stole the continent's lands in the first place; those whose ancestors were enslaved, shipped across the Atlantic, and forced to work the land; and recent immigrants whose arduous, underpaid, exploited labor underpins much food production and processing today.

Land reform will need to be accompanied by an overhaul of the agricultural economy to make it more equitable. In 2023, farmers received more than $200 billion in federal subsidies such as cash payments and crop insurance. Those subsidy programs are the single biggest reason for the extreme lack of both crop diversity and human diversity in US agriculture. The lions' share of payments are directed at a small handful of field crops—corn, soybean, wheat, cotton, rice, and sugar—along with dairy prod-ucts. None of those crops would be economically viable for most farmers to grow without support from government programs; con-versely, a diverse range of other crops now grown on far-smaller acreages would be more profitable and more widely grown if they did receive subsidies. But major growers of the major crops have disproportionately heavy political clout and gobble up almost all the money.

The 10 percent of US farms with the largest crop sales (over-whelmingly white-owned) receive more than two-thirds of federal subsidies. On average, therefore, Black, Hispanic, and female farmers receive lower payments than their white or male counterparts. That's largely because their farms are smaller, with consequently smaller sales, and they tend to grow a more diverse array of crops, including many that aren't eligible for subsidies.[202]

Cropland ownership is at least as unequal as allocation of federal farm subsidies. The distribution of US farmland is wildly disproportionate to the nation's demography. For example, the Congressional Research Service (CRS) reported in 2021 that 96 percent of US farmers are white and that since 1900, the proportion of farmers who are Black declined from 13 percent in 1900 to only 1.4 percent in 2017.[203] Diversification of farming, sorely needed, would have beneficial effects throughout society. And Washington has utterly failed to compensate Black, Indigenous, or Latino farmers for the centuries of discrimination that created this wild disproportion. The American Rescue Plan, a $2 trillion bill passed by Congress in 2021 as part of the recovery from the Covid-19 pandemic, allocated $4 billion to debt relief for Black farmers. But after white farmers filed a lawsuit claiming that the law would make *them* victims of discrimination, Congress revoked the funding for Black farmers and passed a different bill that provided only one-twentieth as much debt-relief funding and was open to farmers of all racial backgrounds.[204]

A path must be cleared for Black farmers to once again play a prominent role in US agriculture. Meanwhile, immigrant farmworkers, freed by breakup of the sprawling, irrational, cruel vegetable-growing operations that employ them today, will have the opportunity to raise crops on farms of their own. And, crucially, Indigenous people will once again be free to live on and care for large tracts of the land that were stolen from them.

The national Indigenous organization NDN Collective is among those in the forefront of North America's Land Back movement. The group stresses that "Land Back," while it's meant literally—that is, Indigenous people's lands must be returned—is also intended more broadly as "a synonym for decolonialization and dismantling white supremacy." That's because, in NDN's words, "our symptoms of oppression are all rooted in our loss of land."[205] In discussing the various mechanisms for returning tracts of land to Indigenous care—co-stewardship agreements, conservation easements, sale, or a straightforward hand-back to the tribes—legal scholars and tribal members Vanessa Racehorse and Anna Hohag argue for the straight handover:

> While the preferred method used by federal and state governments . . . has been to extend opportunities for tribal co-management, this is not enough to curb the urgency of the impending climate disaster, the effects of which have been, and will continue to be, felt first and foremost by Indigenous peoples. It is time for Land Back. There is no clearer argument for Land Back than to prevent irreparable harm to the planet—a cause that is unquestionably in the greatest good for all people.[206]

The broader meanings of "Land Back" are highly compatible with those of degrowth. Asked in an interview with the publication *Civil Eats*, "How do you reach a balance between doing local work and sharing solutions among tribes and outside Indigenous communities so that the projects and their impacts are scalable and have more impact?" Jade Begay, NDN's director of climate justice, was blunt:

> This word "scalable" is one I hear all the time in the climate space, and I really have to turn this word on its own head and reframe it. Because we don't need scale. The term "scalable" is rooted in capitalism. And what we're needing right now is anti-capitalist solutions to the climate crisis. When we show up in spaces like COP26, "scalable solutions" is what the majority of industry, tech, and governments are trying to do. It's a buzzword. What we do is look at solutions in terms of an Indigenous community, nation, or tribe. Each of these communities has its own traditional ecological knowledge to inform its own solutions. We're not interested

in blanket solutions that are meant to keep production or consumption going at the rate that it has been going at.[207]

Recent years have seen some tentative moves toward Land Back. Though they're mostly baby steps, they are encouraging. Through Joint Secretarial Order No. 3403, signed in 2021, the Interior and Agriculture Departments finally began returning at least a small portion of Native lands through co-stewardship agreements or other means. As of early 2025, the tribes had entered into more than four hundred cooperative agreements with federal agencies to comanage and co-steward their ancestral lands.[208] It was nothing close to transformative change, but it was a step in the right direction.

Even before 2021, change was happening. An eighteen-thousand-acre stretch of Western Montana, known up to that time as the National Bison Range, was restored to its original inhabitants, the Confederated Salish and Kootenai Tribes, as a federal trust ownership in 2019. A brief history on the website of what is now known as the CSKT Bison Range explains that the original range

> was established in 1908 by the US Federal Government in the middle of our treaty-reserved home, the Flathead Indian Reservation, on land taken without our consent. The bison living on the Bison Range also are descendants of the free-ranging Reservation herd started by Tribal members in the 1800s, when plains bison were near extinction.[209]

Now, after more than 125 years in exile, the Salish and Kootenai Tribes can once again live in and with those twenty-eight square miles of landscape, among the bison, fish, edible and medicinal plants, and countless other species integral to that ecosystem. They're managing the bison within the range as well as helping manage herds from Yellowstone National Park that graze on US

Forest Service land. Tom McDonald, a Fish and Wildlife Service manager and tribal member, told *Yale E360*, "We treat the buffalo with less stress, and handle them with more respect." That, he said, "was a paradigm shift from what we call the ranching rodeo type mentality here, where they were storming the buffalo and stampeding animals. It was really kind of a violent, stressful affair."[210]

Tribes in the state of California have been taking over management of their traditional lands. A group of five tribes is protecting more than two hundred miles of coastal territory that has been degraded by climate change and economic activities. Following a dam removal, they've been monitoring salmon populations, testing for toxins in shellfish, and carrying out other actions.[211] In 2022, the state returned the Butte Creek Ecological Preserve to the Mechoopda Indian Tribe of Chico Rancheria. Butte Creek is home to an array of important species, including the western pond turtle, yellow-legged frog, and salmon. It's the spawning grounds for the largest population of Central Valley spring-run chinook, a threatened species of salmon.[212]

State governments in California, Kansas, Virginia, New York, and Maine have entered into other land co-stewardship agreements with Native tribes.[213] The Yurok, who have been living on the banks of the Klamath River in the northwest corner of California for ten thousand years, are getting back portions of their ancestral lands stolen long ago. (Shamefully, though, the government is making them pay for the land.) To date, they have acquired eighty thousand acres within which four salmon-spawning streams flow through a vast area formerly owned and abused by a timber company. Salmon has been the tribe's main food source for millennia, and members will be working to restore their habitat while continuing to gather terrestrial foods, now across a far wider range.[214]

Across the continent in Maine, five tribes of the Wabanaki Confederacy, along with a group of land trusts, are working to gain access to ancestral lands throughout the state, including a one-hundred-fifty-acre island that the state of Maine seized from them in the nineteenth century. There, they can once again practice hunting and gathering.[215]

Sadly, even this slow, steady progress ground to a halt in January 2025, when the Secretary of the Interior implemented a Trump administration executive order to drop all programs related to equity or environmental justice. [216]

CROSS INJUSTICE OFF THE SHOPPING LIST

Degrowth in crop growing may sound like an oxymoron, but it's necessary if we're to have a rational food system. The agricultural sector will be able to produce enough of the kinds of foods that can nourish and sustain everyone, regardless of income, while simultaneously phasing out wasteful, superfluous, and destructive food production. First to be eliminated should be biofuel production, factory farming of animals, and the routine abuse of soils, waters, and the atmosphere.

In sectors with high labor demand, the exploitation and abuse of farmworkers and food-processing workers, including those in the meat industry, will end. Food will be produced at a humane, ecologically supportable pace by worker-owned cooperatives paying a living wage. In degrowth writings, there's a lot said about producing as much food as possible—especially fresh produce and fruit and nut trees—as close as possible to those who eat them. Backyards, community gardens, community-supported agriculture, and other small-scale production feature prominently.[217]

Because vegetables and fruits together occupy less than 5 percent of national food-growing land, more of that production can easily be moved into urban and suburban areas (especially after most of the parking lots have been busted up and hauled off), even as garden-scale production continues in rural areas and small towns. In regions with cold winters, it will be seasons, not space, that limit how much can be produced and when. For both ecological and humanitarian reasons, the future cannot continue featuring centralized, industrial-scale vegetable production in warm climates and the industrial preservation and transportation systems required to distribute fresh produce continent wide. In much of North America, therefore, the range of fresh produce available will necessarily be smaller and different in wintertime. That doesn't mean that Christmas dinner will be vegetable free. Fruits and vegetables will just have to be produced locally in late spring, summer, and fall, with local, cooperative canning, drying, and other forms of preservation employed to boost availability in winter and spring.

Saying goodbye to fresh tomatoes and strawberries in winter, sad as it might be for many Americans, will be necessary for the collective good if we strive to become a more moral and just nation. Migrant workers in fruit and vegetable production on the West Coast and in Arizona, Texas, Florida, New Jersey, and other states suffer extreme labor exploitation and inhumane living and working conditions, as do the field workers in Mexico and the rest of Latin America who produce food for export to the North. In the United States, migrant workers were among the groups hit hardest by the coronavirus. From the beginning, they were designated "essential workers," but they got little of the respect and gratitude that should have come with such a designation. Little effort was made to limit their exposure to the virus, and they were pushed harder than ever.[218]

By 2022, average farm-labor pay had increased somewhat, thanks in part to minimum wages set by the federal government in the visa program under which immigrants are hired. But even with their improved wages, immigrant farmworkers were being paid far less than US workers.[219] With degrowth, it will be time to jettison the wage-labor system in vegetable production and let worker-owned cooperatives take over.

NATURE IN SURPLUS

In a future society where most people participate in their own food provisioning, where rural areas are repopulating, and in urban areas where, with paved area shrinking dramatically, and with green spaces bursting out all around, people would have the pleasure of saying good riddance to one more, very important ill in today's society: separation from the rest of nature.

In cities and suburbs freed of automobile supremacy, expanses of concrete and asphalt no longer needed for driving and storing motor vehicles are replaced by diverse stands of food-producing plants. Others become home to native plant species that support pollinator populations. Still others become home to forests of fruit and nut trees. Yet more space remains available for fostering new green areas and expanding old ones. With suburban sprawl reversed, population centers become surrounded by mosaics of forests, prairies, and cropped lands dotted with villages that are connected to the city by public transport and foot and bike trails. Beyond the horizon, on the landscapes no longer known as "fly-over country," a century and a half of rural-to-urban migration shifts into reverse. With plenty of space and plenty of work to do, the now more numerous, smaller farms and reviving small towns repopulate. Wildlife, never having left but now under a far smaller threat of suddenly becoming roadkill, thrive. After a perennial conversion of agriculture, streams run clear again.

All the above would go a long way toward narrowing the distance and healing the rift between humans and nature that plagues technological society. The commonly cited finding that Americans spend an average of 90 percent of our lives indoors appears to be well-founded, and it has stayed fairly constant since the mid-twentieth century (not coincidentally, during the rise to dominance of air-conditioning).[220] In his 2008 book *The Last Child in the Woods*, Richard Louv coined the nonmedical term "nature deficit disorder" to capture technological societies' severe paucity of experience in the nonhuman-made world.[221]

Before and since the publication of Louv's book, stacks of peer-reviewed publications have demonstrated the wide-ranging benefits of spending time in more natural environments. The advantages include reduced rates of vitamin D deficiency, high blood pressure, diabetes, obesity, mental health problems, attention deficit disorder, stress, mood disorders, and even premature births. Children enjoying larger doses of nature showed higher cognitive functioning and had higher test scores. In some studies, children who simply had a view of green spaces from their apartment building exhibited less procrastination, more self-discipline, better social relations, and less violence.[222]

There are several paths through which natural ecosystems benefit human health: strengthening of immune systems when we repeatedly encounter outdoor microorganisms; our tendency to exercise and socialize more when we're in more natural environments; and the reduced amounts of air, chemical, and noise pollution that we're exposed to in greener spaces.[223]

The group NatureQuant assigns a rating that it calls a "NatureScore" to each of the more than eighty-four thousand US Census tracts. Scoring of each neighborhood-sized area on this zero-to-one hundred scale (with higher numbers greener) was based on satellite imagery and variables such as air and noise pollution, park (not parking) space, surface water, and tree cover. It should come as no surprise that the surveys found green space and its many benefits to be distributed inequitably. The 20 percent of tracts with highest median incomes had an average "NatureScore" of 64, while the economically poorest 20 percent of tracts averaged only 50. Racial disparities were even greater. The "whitest" 20 percent of tracts scored 73, the least white just 45.[224]

Urban food provisioning can be part of a wider movement toward life outdoors as an active part of nature. Among the green urban features that could accomplish that reintegration, the organization Nature Canada lists food-oriented activities such as community gardens, orchards, food forests, and foraging alongside conservation (tree planting, etc.), education (bird watching, citizen science), and recreation (hiking and biking).[225]

There are plenty of examples showing how food systems and connections to nature, both often badly broken in the United States today, could be repaired in urban and suburban areas by creating much more green space. Southeast Atlanta is home to several majority-Black communities as well as the seven-acre Urban Food Forest at Browns Mill—the largest public food forest in the country. Browns Mill has some tilled vegetable gardens, but most of the area is woodland where Atlantans can forage at will from nut trees, fruit trees, berry bushes, and low-growing plants like strawberries, grape vines, herbs, root vegetables, and other plants. A study by US Forest Service and University of Georgia researchers noted that "City-dwelling African Americans have a long history of growing and consuming foods from household gardens" and that one of every five residents they interviewed participates in foraging.[226]

Suggestions for "reconnecting with nature" have traditionally reflected the concept of nature as an entity separate from humanity. In this view, the "cure" is for humans, especially children, to go out and "visit" Mother Nature by, for example, spending time in national parks, wilderness areas, botanical gardens, nature trails, and so forth. Contact with nature is thus viewed as a treatment to be undergone from time to time, not as a part of everyday life. Experiencing ecosystems that are less affected by human impacts is of course beneficial in many ways, and that opportunity should be more readily available to all; however, for many families, travel to experience wild nature is unaffordable. To provide physical and mental benefits like those cited above, a green space need not be on par with Yosemite or the Serengeti. It could be a city park or Atlanta's food forest or a creek running through a prairie in the grazing lands of Kansas, or just a community vegetable garden.

As nature flourished during the 2020 anthropause, people living in places under lockdown were not thriving. Rural people maintained their closer contact with green spaces, and they, especially rural children, fared better than those living in cities. But kids in urban areas who had a home with a backyard, even if it was just a small patch of green where they could go out and play safely, suffered far less stress than those who didn't.[227] With degrowth, access to green space will be a universal right, not a privilege.

NEW OLD LANDSCAPES

One more big step is needed in this imagined transformation of food and farming, but first, let's recall what has been achieved so far in this scenario. We've managed to banish corn and soybean from tens of millions of acres of land. That has freed up vast acreages for expanding both the sown area and diversity of food crops and restoring much of the recovered land to grassland or forest. We've done away with factory farming of everything from cattle to tomatoes and moved vegetable and fruit production closer to where people live. We've abolished megafarms and pursued land reform, creating opportunities for many people across a much more diverse array of communities to become farmers. And Native tribes are reclaiming and restoring the home ecosystems of their ancestors.

The nation in that vision will have come a long way, but left unaddressed is the fundamental problem of agriculture: humanity's ten-thousand-year history of killing off natural ecosystems and replacing them with stands of annual field crops. That's a problem for many reasons. Plant communities in natural ecosystems are almost entirely diverse arrays of perennial species, whereas the grain and root crops and other species that predominate in agriculture are annual species incapable of sustaining robust, diverse soil ecosystems as perennials do. To make matters worse, annual crops, sprouting from seed every season, are usually sown in monocultures—one crop species per field. Intact ecosystems, in contrast, are self-sustaining and run on sunlight, and their diverse plant communities are anchored by extensive, year-round perennial root systems that are essential to sustaining healthy soil ecosystems and preventing erosion. Fields of annuals, by contrast, are not self-sustaining, require energy inputs far beyond what sunlight can provide directly, and must reestablish their root system every season. For months at a time, their soil is either devoid of crop roots or is nourished only by young, sparse, shallow root systems. Because the soil becomes more impoverished with every year that it's tilled, it requires life support in the form of fertilizers and other chemical inputs. But those fixes only further degrade the landscape.

The major food grains, including wheat, rice, oats, dry beans, peanuts, sunflower, and all others, are annual species. They cannot be cultivated without some degree of soil erosion, destruction of belowground ecosystems, and impacts on waterways and groundwater. Therefore, perennial-grain-cropping systems are needed to produce human food while providing some of the ecological protections afforded by natural plant communities. Until recently, domestication and breeding of perennial grain-producing crops had a sporadic history going back to the early years of the Soviet Union. Today, the longest-running perennial grain-breeding programs are led by The Land Institute in Salina, Kansas (where I worked for twenty-five years). More recently, perennial-grain development has been taken up by more than fifty researchers across North America and most other continents. Breeding of perennial wheat continues at The Land Institute, which is also developing a perennial cousin of wheat known as Kernza for pilot production in the US Plains, the upper Midwest, and Europe. Highly productive perennial-rice strains are being grown on tens of thousands of acres in China. Breeding and ecological work continues in countries around the world, with those three crops, plus perennial-food legumes, perennial-grain sorghum, and others under development.

Conversion to largely perennial-grain agriculture would require an ambitious national effort that is sustained over the long term. But the US Congress funds the Department of Agriculture for only five years at a time through successive pieces of legislation popularly known as Farm Bills. In 2009, The Land Institute's cofounder Wes Jackson and then director of Iowa State University's Leopold Center Fred Kirschenmann, released a white paper titled "A 50-Year Farm Bill." The plan proposed that annual field crops be displaced by perennials throughout farm country. The gradual substitution would begin with perennial pasture, forage, and hay crops. Then, after a couple of decades, deployment of newly developed perennial-food crops—cereals, grain legumes, and oilseeds—could begin and gradually cover more land area. Jackson and Kirschenmann foresaw that during the fifty-year transition, the makeup of the agricultural landscape would be flipped from 80 percent annual and 20 percent perennial to 80 percent perennial and 20 percent annual. [228]

The benefits of a transformed agriculture would come gradually. But with degrowth, a future US society will leave behind a host of social and environmental problems. As the fruits of a fifty-year farm bill ripen, farm households will be freed from burdens as old as agriculture itself. The most obvious is soil erosion and degradation, which has brought down civilizations large and small across the globe. In a 2003 essay, Jackson painted a picture of security and peace of mind that could come to a farm family someday when the specter of soil erosion has been left behind:

> Early May: Rain. Three days in a row now, and into the night it kept coming. The south bottom is flooded. Half-Day Creek was over its banks and had spread into the east pasture this morning when Dad moved the yearlings. More rain is in the forecast.
>
> It's calving time. Eight have arrived and there are four more to come. This is the family's concern. It is not soil erosion, or muddy washes or replanting corn after such weather. All of these are memories held in family lore.
>
> The rain is relentless but welcome. The two-year drought is over. Last year's precipitation, less than half the average, has been recouped with rains of the past month.
>
> The drought had not seriously cut last year's yields. Farmers now have mixtures of perennial plants whose varying root architectures are nature's designs to handle drought, absorb water and manage nutrients efficiently.
>
> Some reach ten feet down and more; they use water stored earlier. Now, in a downpour, those roots hold the soil, and everybody in the house this morning knows that the soils on the farm are weatherproof.
>
> It was not always so. They knew of the fifty-year transition their grandparents made on this land. Gone from the rolling Midwestern countryside are the monocultures of annual crops . . . But with the perennial root mixtures anchoring grain crops now, for the first time in ten thousand years, farmers are not forced to roll dice against gravity. They had won rarely. Less than a century before, the farmer who stayed with it was like the gambler who stays at the casino betting against the house. Because wind and rain are the

norm, anyone with annual crops on sloping land lost soil. And no matter how deep, there was only so much good soil to lose.

By Thanksgiving, all of the fields of these modern domestic prairie patches will have been harvested . . . What makes this a different agriculture is not the aboveground drama at harvest. It is the shift of attention over the past century to processes below the land's surface.[229]

River waters running through agricultural landscapes will no longer be a muddy-brown color. Nitrates from fertilizers and other chemical compounds will no longer render farm-country streams, rivers, ponds, and lakes unsafe for swimming; consumption of fish caught in water bodies will no longer be limited for health reasons. Skies in the Great Plains will no longer be darkened by dust storms composed of priceless topsoil swept up from bare fields as they waited to be sown to annual crops. Rural residents of Nebraska, Minnesota, Iowa, Illinois, and other states will no longer look out on fields of corn, soybean, and very little else, stretching to the horizon in summer and fall, and then in winter and spring, vast expanses of bare soil. No longer will our society strain to find ways to dispose of countless tons of hazardous waste from feedlots and other animal-confinement facilities. Farmers and farm laborers will no longer succumb to cancers caused by farm chemicals.[230] Grain farmers will no longer take the income they receive from sale of their harvests (and government subsidies), turn around, and hand most of it over to the companies who sell them seeds, fertilizers, herbicides, pesticides, tractors, combines, and myriad other inputs and equipment.[231] Farm families will no longer be harassed by pipeline-company land grabs. And there will be no more barriers preventing Black, Latino, and Indigenous families, and immigrants of all backgrounds, from becoming—and thriving—as farmers.

Cities, suburbs, and towns will be busy expanding natural green space while producing more of their own food. And they'll no longer be willing to sacrifice land area to the sterile aesthetics of lawn culture, professional landscaping, and parking lots. No longer will urban and suburban dwellers want to spend more than 90 percent of their time indoors or feel disconnected from the soil and the living world.[232] "Food deserts" will exist neither in low-income urban areas nor in small towns in farm country, and boutique food stores will be gone.[233] Supermarkets and home kitchens will no longer be stocked with wide varieties of corporate packages containing extremely narrow ranges of food sources. And quack dieting culture, phony dietary supplements, and overprocessed foods will recede further and further into the past.

V.

Take My Job—Please!

In 2013, the anthropologist and writer David Graeber penned an essay titled, "On the Phenomenon of Bullshit Jobs," for the British magazine *Strike*.[234] In it, he argued that technological advances have so greatly increased workers' productivity that twenty-first-century Global North economies, in which a large majority of adults are employed, face a problem. The number of person-hours of paid work time far exceeds what is necessary to produce the quantities of basic goods and services required by the population, so to fill those hours, he wrote, employers have created millions of "bullshit jobs" that benefit neither their companies nor society at large.

Graeber followed up five years later with a book titled *Bullshit Jobs: A Theory*. In it, he defined his subject more precisely as a "form of employment that is so completely pointless, unnecessary, or pernicious that even the employee cannot justify its existence even though, as part of the conditions of employment, the employee feels obliged to pretend that this is not the case"[235] As he hastened to point out, this doesn't mean there are millions upon millions of working people enjoying cushy, easygoing workdays while raking in a nice paycheck every month. Ample evidence shows that most adults are happiest when they have purposeful, fulfilling work to do, whether it's paid or unpaid.

They don't enjoy spending countless hours doing something meaningless and unnecessary, whatever the wage or salary—especially if they are also compelled to pretend that their work is important to society.

At the time he wrote the original essay, Graeber guessed that roughly 20 percent of the working population of Britain, where he lived, worked in bullshit jobs (and, by definition, they knew they did). After the article went viral in his country and then flashed around the world, the survey firm YouGov conducted a poll of UK residents asking questions based on his essay, such as "Does your job make a meaningful contribution to the world?" To that question, 37 percent of respondents answered no—almost double Graeber's private guess—and only half answered in the affirmative. Responses to a subsequent poll in the Netherlands found 40 percent of workers believing that "their jobs had no good reason to exist"—precisely doubling Graber's original guess for the UK. Hundreds of responses from around the world, posted to social media or comments sections or written directly to Graeber, confirmed a deep dissatisfaction in the workforce and spurred him to write the book. Its preface ended with this paragraph:

> We have become a civilization based on work—not even "productive work" but work as an end and meaning in itself. We have come to believe that men and women who do not work harder than they wish at jobs they do not particularly enjoy are bad people unworthy of love, care, or assistance from their communities. It is as if we have collectively acquiesced to our own enslavement. The main political reaction to our awareness that half the time we are engaged in utterly meaningless or even counterproductive activities—usually under the orders of a person we dislike—is to rankle with resentment over the fact there might be others out there who

are not in the same trap. As a result, hatred, resentment, and suspicion have become the glue that holds society together. This is a disastrous state of affairs. I wish it to end.[236]

JOBS TO LEAVE BEHIND

Graeber's thesis and the poll numbers supporting it suggest that today's affluent societies could sustain themselves with far fewer hours of paid work than prevail today. The same will be true in a degrowing society that's using less energy and material resources and producing less monetary wealth; jobs that use up resources and burden our planet without improving the lives of either employees or society at large will be out. Other, more socially beneficial positions will be created. The net result is a lighter and more equitably distributed total workload. And a much larger share of the population will have fulfilling employment that benefits society while also having more free time for self provisioning and many other activities.

There are countless people whose job is not of the bullshit variety, by Graeber's definition. Some of them like their jobs and believe their work is necessary and makes a positive contribution. (I had such a job before I retired in 2025.) Others believe their job makes a positive contribution to society, but they simply dislike their work or workplace environment.

Whether or not they view it as pointless, people can be dissatisfied with their job for a host of other reasons. A massive 2019 survey of seventeen thousand workers in nineteen different industries uncovered several sources of so much workplace unhappiness. Almost two-thirds of respondents said their supervisor rarely "gives support when things get tough," while three-quarters said their fellow employees are rarely "held accountable for their work, regardless of position" and that their work environment is "always or often . . . overly focused on trivial activities or has overly bureaucratic company policies." [237]

Large majorities of workers responded that they always, often, or sometimes experience the following:

- o Stress from job affects relationships with family and friends: 81 percent

- o Tend to work alone because of an unhelpful or hostile environment: 63 percent

- o Speak poorly about company to colleagues or people outside work: 71 percent

- o Thinking about or looking for a different job: 71 percent

- o Spend thirty-one to forty hours per week distracted from work: 77 percent

Meanwhile, in a different 2019 poll, this one by Gallup, only 40 percent of workers characterized their employment as "good," with the rest responding "mediocre" or "bad." Notably, the frequency of the response "bad" was almost twice as high among Black women as among respondents in general. [238]

I recently did a web search for the sentence "I hate my job" and these were six of the top seven article links: "'I Hate My Job': How to Cope When You Feel This Way"; "I Hate My Job!—What to Do Now (Follow These Ten Steps)"; "What Do I Do if I Hate My Job and Feel Clueless?"; "I Hate My Job! What You Can Do When You Hate Work"; "I Hate My Job: What To Do When You Can't Take It Anymore"; and "'I Hate My Job!' Here's What You Can Do (Before Quitting)." [239]

Whether or not the ample supply of "I Hate My Job" advice

reflects widespread dissatisfaction, no employee should be compelled to adapt to an intolerable environment for half or more of their waking hours. On the long list of things that most people would be happy to say goodbye to if there's a societal shift into degrowth mode, crappy jobs might end up close to the top.

The burdens of inadequate pay, long or irregular work hours, night work, exploitation, and a hostile work environment are also distributed inequitably. On average, as we all know, American men earn more than women for similar work. White women earn more than Black, Latina, Indigenous, and other women. Native-born workers earn a lot more than immigrants. White men earn more than every other group. And typically, the lower one's pay and job status, the worse one's working conditions and other aspects of the job.[240] There are also big disparities in the hours of unpaid labor performed at home—the kind of work without which US society simply couldn't function. Women under forty-five spend almost two and a half times as many hours in unpaid household and care work as men in that age range.[241] Tackling such disparities is a high priority for degrowth.

THE STARTING LINE: SHORTER WORKWEEKS

One of the elements most prominent in degrowth visions is reduced working hours, or what Timothée Parrique prefers to call "employment time reduction." He suggests that framing because, he says, what needs to be reduced is the number of hours spent in *wage* labor, especially labor that's "alienating, exploitative, and divisive."[242]

To people who declare "I hate my job," every workday is too long. Even many workers who like their jobs say they'd prefer to spend less time at work, allowing more time for the rest of life.[243] There's no good reason for an ironclad minimum forty-hour week; the owning and investing class simply ordains it. As the University of Leeds economics professor David Spencer puts it, "The continued force of consumerism has acted as a prop to the work ethic. Advertising and product innovation have created a culture where longer hours have been accepted as normal, even while they have inhibited the freedom of workers to live well."[244]

To be sure, marginally shorter workdays or workweeks alone won't cure all the ills of being employed in a capitalist economy. But recent research on time spent on the job versus off the job can help us envision the better quality of life to be found in a degrowth society that has found ways to, as researchers at the New Economics Foundation have put it, "break the habit of living to work, working to earn, and earning to consume."[245]

A 2022 experiment involving 360,000 workers in the United States, Ireland, Australia, and New Zealand found that people working thirty-two-hour weeks for forty hours' pay suffered less stress and fatigue, were happier, and enjoyed better health. A smaller survey in the United Kingdom found similar results. Many respondents said they were getting "more done in less time" and had felt less physical and mental stress and suffered less from burnout. With a 20 percent reduction in days worked per week, their hours spent commuting by car fell even further, by 27 percent.[246] Four-day-week workers in Ireland told researchers they were spending more time volunteering in their communities, engaging in exercise, walking or cycling instead of driving, and participating in environmental education. Perhaps most remarkably, rates of sleep deprivation fell from 34 to 9 percent.[247]

What did people report having done with their additional day off each week? Mostly errands and housework, indoors and out. That left weekends free for more recreation, hobbies, and family activities. Less of the week was spent in "time-wasting meetings." Stress and burnout were reduced. Fully 92 percent of workers in the experiments said they wanted the four-day week to become permanent.[248]

The ecological economist Juliet Schor reports that 95 percent of company heads who tested the four-day, eight-hour-per-day week did not want to go back. An eye-popping 83 percent of workers under forty-two years old said the four-day week with five days' pay should be standard. Short weeks are still far from common, of course; for example, only 12 percent of US workers have had the opportunity to choose a four-day week.[249] But as more employers experiment with a shorter week, the trend could become largely irreversible. An impressive 70 percent of thirty-two-hour workers in a survey in Spain said that if told they had to go back to a forty-hour week, they'd refuse and quit unless they received steep pay increases ranging between 10 and 50 percent.[250]

That said, the consequences of the four-day week, beneficial as they are, fall well short of workplace transformation. Unlike the adoption of the eight-hour day, won by workers across the globe through decades of organized struggle, most of the more recent experiments with a four-day week were conceived and organized by outside advocacy groups. A review of thirty-one four-day-week studies published since the 1970s found that on the positive side, shorter working hours helped improve workers' morale and job satisfaction and decreased their time spent in the ordeal of commuting. But there were also negative effects: complications with scheduling, increases in workload on the first day back from a long weekend, childcare woes, and more absenteeism. As a result, concluded the review's author, "performance management, monitoring, and productivity measures were intensified. Pro-social and collective interests evident in labor-led campaigns were absent."[251] Authors of one of those thirty-one studies, published in 2022, concluded that

> in order to gain some freedom *from* work, employees were prepared to relinquish some freedom *in* work. Hence, this version of the four-day week joins a long line of other

work-life flexibility initiatives that promise freedom and yet
ultimately serve to strengthen employees' investment in capitalist work and organizations.[252] (Emphasis added.)

A workweek of around thirty hours, increasingly embraced by
both workers and management today, was raised as a demand by
US autoworkers as long ago as the 1950s. But seventy years later,
it remains a nonstarter in that industry. When the United Auto
Workers (UAW) union won a landmark new contract from the
Big Three automakers in December 2023, the workers got much
of what they were aiming for, but shorter hours remained out of
reach. Chris Isidore of CNN reported that none of the UAW's
bargaining goals "was more ambitious, or gained less traction in
negotiations, than the idea of a four-day, thirty-two-hour work
week" with no pay cut.[253]

In 2020–21, as the Covid-19 pandemic gripped the nation,
flexible-working hours, remote work, and other adjustments
were adopted to protect many white-collar employees, but
workers who manufacture, process, handle, haul, or service
industrial or agricultural goods were often out of luck. Many,
like those in the meatpacking industry, suffered even more
severe exploitation than usual, topped off by greater Covid risk.

Owners and managers are happy to shorten hours when total
production can be sustained or even increased through speedups
and other super-exploitation. But workers on a Ford assembly
line or in an Amazon warehouse are already working flat out.
Increasing their hourly output by 25 percent to sustain their forty-hour output in a thirty-two-hour week would be out of the
question.

So it is that among seventy companies offering four-day workweeks with five-day pay as of 2022, most, according to *Fortune*,
were in the start-up and digital-tech sectors. Here's a random
assortment from that list: Kickstarter, Halftone Digital, eFileCab-

inet, Charlie Hustle, Buffer, and Founders Club. There were also a few nonprofits (including Common Future) and a local government (Boulder County, Colorado). The list contained no firms engaged in manufacturing, construction, or resources.[254]

GO DOWN TO TWENTY HOURS FOR A
WHOLE NEW BALL GAME

Situated well within the boundaries of business as usual, the thirty-two-hour workweek falls far short of being transformative. It's embraced by management only in industries and situations where it will not reduce total work output or profits. And it doesn't change the nature of work. Employees like it, but not necessarily because it makes their jobs less onerous or more rewarding; they like it because they spend more time *away* from their jobs. A much deeper reduction in paid-work time is both possible and desirable, but it would require a dramatic, even revolutionary, reworking of economics and governmental function.

The Great Recession prompted widespread discussion about even shorter work hours. In the years following the 2008 economic crash, the UK-based New Economics Foundation launched a campaign for a twenty-one-hour workweek. They predicted that unlike the four-day week, a "'normal' working week of twenty-one hours could help to address a range of urgent, interlinked problems: overwork, unemployment, over-consumption, high carbon emissions, low well-being, entrenched inequalities, and the lack of time to live sustainably, to care for each other, and simply to enjoy life." They went on to stress how the difference between twenty-one and thirty-two hours goes far beyond the sheer number of hours:

> Of course, a four-day week would be a significant step in the right direction. But it would leave undisturbed the current

norm in which everyday life is structured around delineated hours of paid work, shaped by its overriding demands, and imbued with associated values. A twenty-one-hour week, or its equivalent in hours spread across a month or year, over-turns that scenario. It forces us to consider a different set of relationships between time, money, and consumption, as well as how these new co-ordinates might affect the dis-tribution of power between people and groups, what really matters for human well-being, and how we can carve out a *sustainable* future. [255]

Why twenty-one hours and not twenty, twenty-four, or some other roundish number? The twenty-one-hour figure wasn't based, as one might guess, on a three-day, seven-hour-per-day workweek. Rather, it happened to be the average number of hours per week that adult Britishers were working at the time. That figure lumped together millions who were severely overworked with millions who were underemployed or had no paid job at all and needed one. An effort to make twenty-one hours the standard workweek, distribute the national workload more evenly across the population, and pro-vide for individual flexibility in employment time could result in a UK working population that provides the same total number of person-hours as the current workforce. It could also leave workers and their families free to enjoy a better all-around quality of life.

Neither capitalists nor their growth economy would tolerate a twenty-hour workweek. [256] Degrowth advocates maintain that a non-capitalist economy, by contrast, could function well on an average workweek of twenty hours and, over time, maybe even shorten it further. Such an economy will be unencumbered by requirements to overproduce, oversell, and overconsume but must be fully equipped with elements that are lacking in most economies today: employee control, job sharing, extended-care leave, and a more even distribution of work time and earnings throughout the population, especially between the sexes. Further ensuring economic fairness would require income guarantees, steeply progressive taxation, higher minimum wages, universal health care and childcare, and a high societal value attached to care work that happens outside the market. The New Economics Foundation's report envisioned that a twenty-one-hour workweek could help make such policies possible, because it would provide more time to be an active citizen: "We could do things with and for each other that we might otherwise have to buy—exchanging knowledge and skills, running errands and caring in ways that have been tried and tested for generations through mutual aid schemes and timebanks."[257]

Less time spent in paid employment would boost a degrowth society's efforts to meet its goal of deeply reducing ecological impact. Analysis of work, income, and environmental impact data from twenty-nine European countries showed that economies in which people spend fewer hours doing paid work tend to do less ecological damage. It has long been known that greater affluence produces higher carbon dioxide emissions and more severe ecological impact, and the European study confirmed that. But it also showed that the shorter a given country's average workweek, the lower its emissions and the smaller its ecological impact per person, whatever the country's relative wealth or population.[258]

VETERANS OF BULLSHIT JOBS SPEAK

There's plenty of room in affluent economies for reducing paid-work hours. A society that's extracting fewer energy and material resources, manufacturing less of what's not needed, traveling less, conducting less commerce, and making fewer monetary transactions will have less need for labor in the industries that deal with those activities. The reduction in labor demand will be partially offset by increased demand elsewhere. That's because as fossil fuels are steadily phased out, the quantities of goods and services produced per worker will decrease. The economy will also require increased human power in areas such as food production and manufacturing. However, there will still be many opportunities for reducing employment time.

Among those targets of opportunity, the economic sector dealing with finance, insurance, real estate, rental, and leasing—which makes up 20 percent of the US economy, almost twice as much as manufacturing—will have lots of capacity for downsizing, and that will make room for both more useful work and more leisure.[259] This sector (often abbreviated as FIRE) is devoted, in David Graeber's words, to "creating, playing around with, and destroying large amounts of money" and will have little purpose in an economy in which money is not paramount.[260]

Graeber received more than 250 stories from people across the world describing their experiences in paradigmatic bullshit jobs. Most of those jobs were in white-collar sectors, including FIRE, but some were in other types of work. The following are brief summaries of a few of the stories sent to Graeber.[261]

Hannibal was a digital consultant for pharmaceutical companies' marketing departments. His work was writing reports, he said, "with titles like *How to Improve Engagement Among Key Digital Health Care Stakeholders*. It is pure, unadulterated bullshit."

Betsy's job was to "coordinate leisure activities in a care home," which mainly involved interviewing residents and filling out forms with their recreation preferences. The forms were gathered into binders, the information was typed into a computer, and neither set of survey results was looked at again.

Judy had a human resources position in an engineering firm. Her job existed "only because the HR Specialist was lazy and didn't want to leave his desk." She never had more than an hour and a half of work to do in a day.

Ben was in middle management: "Ten people work for me, but from what I can tell, they can all do the work without my oversight. My only function is to hand them work, which I suppose the people that actually generate the work could do themselves." Alphonso's job—coordinating a team of five translators—was similarly superfluous, since "the team is perfectly capable of managing itself: they are trained in all the tools they need to use and they can, of course, manage their time and tasks."

Greta was a receptionist at a Dutch publishing company, where her assignment was to answer a phone that rang only once or twice a day. Therefore, she said, the company assigned her two additional tasks: keeping a candy dish filled with mints and keeping a grandfather clock wound.

Nigel was hired by a company that had a contract to "scan the application forms for hundreds of thousands of company loyalty cards." Each form had to be checked for errors three times by a team of "Data Perfecters," of whom Nigel was one. He told Graeber that "the time seemed to pass quickly, like some kind of near-death experience. There was something about the sheer purity of the social uselessness of this job, combined with the crippling austerity of the process, that united the Data Perfecters."

Tom was in advertising. He told Graeber, "Supply has far outpaced demand in most industries, so now it is demand that is manufactured. My job is a combination of manufacturing demand and then exaggerating the usefulness of the products sold to fix it."

Tom's is among a class of jobs that, Graeber writes, "have an aggressive element" and would be wholly unnecessary in a rational economy: "The most obvious example of this are national armed forces. Countries need armies only because other countries have armies. If no one had an army, armies would not be needed. But the same can be said of most lobbyists, PR specialists, telemarketers, and corporate lawyers."

Recall Graeber's estimate that as many as 40 percent of

employees in rich economies occupy bullshit jobs. In a future economy, they can instead perform necessary, fulfilling work that will take up fewer hours of the day than their old jobs did, leaving time for everything that's not a paid job.

There are lots of other work roles that, while they may not qualify for Graeber's definition of a "bullshit job," are ones that will vanish as we move into degrowth mode. And we can wave them a collective, cheery goodbye. Many occupations would be declared obsolete because they are inherently exploitative, harmful to society at large, or both. Chapters II through IV suggest several areas that are ripe for deep job cuts. Examples are advertising, car sales, industrial meatpacking, and highway construction.

Wasteful, harmful employment even converges with light pollution, our old nemesis from chapter II, in the realm of night shift work. With undesirable industries on their way toward elimination and employees everywhere able to cut their working hours in half, we can finally achieve a deep reduction in night jobs, which have long been known as a serious occupational threat to human well-being. For example, mental and physical health effects of nighttime lighting were first studied among people working the graveyard shift. That's no small population; more than 15 percent of US workers are on night schedules, either full- or part-time.[262]

Night shift work and its risks are especially common in the agriculture- and food-processing sector, where a disproportionate share of workers are also immigrants and/or people of color. Disproportionate numbers of nighttime workers are also younger on average than those who work the day shift—sometimes even too young to work legally. The Labor Department reported in 2023 that the poultry giants Perdue Farms and Tyson Foods had been employing children as young as thirteen years old to clean some of their processing plants in the hours after midnight. According to NPR, "Middle and high school-aged children made up about a third of the overnight shifts at the plant—handling acid and pressure hoses to wash away blood and meat scraps from industrial machines." The previous year, a lawsuit filed by the Labor Department against a company that hired more than thirty children to clean slaughterhouses and meatpacking houses in Minnesota and Nebraska alleges that "These children . . . were [illegally] employed to perform hazardous work cleaning industrial power-driven slaughtering and meat processing equipment on the kill floors . . . in the middle of the night." The FOX 9 TV station in the Twin Cities reported that the thirteen- to seventeen-year olds were made to operate dangerous power equipment using caustic chemicals and that "Several children in Nebraska, including a thirteen-year-old, reported suffering serious chemical burns while on the job."[263]

Traumatic experiences such as these can inflict lasting mental and physical damage on children, especially when it all happens in the middle of the night under the blueish white glare of industrial lighting. Recall that prolonged exposure to artificial light at night is associated with sleep disorders, cardiovascular problems, diabetes, obesity, depression, and suicide. Night shift work is also associated with increased risk of breast, colorectal, lung, prostate, bladder, and pancreatic cancers.[264]

With degrowth, we can expect night shift work to be limited to situations where there is no choice, as in hospitals. With much less industrial meat production, there will certainly be no need for night work in that industry. And the cruelty of subjecting children to wage labor, which should have been eliminated not just in law but in effect long ago, will finally be behind us.

CONSUMPTION THAT'S "SIMPLIFIED, CLEAN, LESS STIGMATIZED, GUILT FREE"

There are many occupations, including some of the above, that wouldn't exist at all were it not for problems created by the everyday

workings of the economy.[265] Relevant to our purposes is an occupation known as "decluttering consultant," created in response to
the problem of material overproduction and overconsumption.
This line of work emerged in the United States in the early 2000s.
In a 2019 article critical of the decluttering phenomenon, writer
Susie Khamis laid out the problem that the industry is ostensibly
addressing: "For many consumers, at least one consequence of
advanced capitalism is of growing concern: the accumulation and
storage of ever more stuff—clothes, furniture, and the dizzying, disposable ephemera of mindless consumption."[266]

A society considering this situation logically would act in a collective way to simply stop producing excess material stuff. But a
rich capitalist economy behaves neither collectively nor logically.
Instead, in this case, it creates a whole new vocation branded as
"professional organizer" or, less formally, "declutterer." Here, from
a study published in 2007, is a description of one firm's methods:

> The professional organizer in our study introduced a partic-
> ular system of organizing homes and offices. The slogan of
> this system is *"See it, Map it, Do it,"* which is a three-stage
> system that helps their clients in various ways from identi-
> fying the problems of cluttering, to organizing their material
> worlds, and further to launching the philosophy of a system
> to enrich their daily lives spiritually. Photos and pictures were
> used in order to help the clients envision their ideal material
> life with less clutter and chaos. They also help clients envision
> their ideals for the flow of materials within their homes. After
> sorting their possessions, those judged as not important to
> their lives are then discarded. Boxes, shelves, and files are used
> to house retained possessions in a more organized fashion.[267]

Professional decluttering doesn't appear to be a bullshit job, since
those performing it believe in the value of their efforts. They told

researchers that they bring the "gift of professional organizing skills and insights to their clients' material and spiritual life." They said 90 percent of the US population "lacks organizing skills," and they wanted to help their clients "enjoy an organized, simplified, clean, less polluted, contaminated and stigmatized, and more guilt-free life."[268] In other words, the decluttering industry helps its customers live more comfortably within an economy built on a cycle of over-production and overconsumption rather than push against it.

Decluttering may have reached peak buzz with Marie Kondo's book *The Life-Changing Magic of Tidying Up*. Published in Japan in 2010 and in North America four years later, it became a *New York Times* number-one bestseller and was declared "one of the most influential books of the decade" by CNN. In 2019, an original Net-flix series titled *Tidying Up with Marie Kondo* became a hit.

Susie Khamis writes that the book and series boiled the declut-tering process down to its essence: "For Kondo, when deciding whether an item should stay or go, only one question should determine the answer: 'Does this spark joy?' From there, she says, 'Imagine yourself living in a space that contains only things that spark joy. Isn't that the lifestyle you dream of?'"[269]

Indeed, argues Khamis, "Joy is exactly what consumers have been trained to expect from their possessions." That's why in the broader culture, grassroots movements for reducing material pro-duction and consumption such as voluntary simplicity, sustainable consumption, frugality, and anarchism are sidelined—despite the fact that, in Khamis's words, those movements are characterized by "vision, empathy, and generosity of spirit"—in favor of a thriving market for decluttering advice and professional organizing.

I mean no offense to professional declutterers or others who work in jobs that are designed to address harms caused by the growth economy. Until that economy is left behind, we'll still be living in it and still need to make a living. But decluttering, like many other services, pumps out a cloud of commodification that obscures the greater value of unpaid care work and other domestic activities. It's an old story. The market creates a plague, in this case, material excess, and makes it difficult for households to deal with it—by requiring long hours away from home, at a job—and then swoops in to offer a "solution." In that old story, it's time to turn to the page that reads "The End."

Maintaining and repairing one's consumer goods also used to be a mostly do-it-yourself activity, at least up to the point when a professional must be called upon. But those activities are now being usurped by the product manufacturers. In response, the degrowth movement's call for decommodifying labor, reducing resource use and waste production, and resisting corporate control emphasizes the importance of open-source knowledge, tool sharing, the "right to repair," and other forms of pushback against corporate control.

Manufacturers also pump up the demand for their products through "planned obsolescence": the intentional design of products to wear out or go out of style within a short time so that they must be replaced by new ones. Faced with planned obsolescence, resourceful consumers can often lengthen the lives of products by refurbishing them. But in recent decades, companies have been snatching away even that DIY recourse. They're tightly restricting access to software needed for maintenance and repair (especially in the auto industry), designing products to require boutique repair tools that only the manufacturer can sell you, barring free access to manuals and other information, monopolizing replacement parts, and even making it difficult or impossible for the customer to open a product up and gain access to its workings. From electronic devices to cars to farm machinery, manufacturers grab excess profits by exerting long-term control over products that people thought they had fully paid for and owned.[270]

According to researchers associated with the Europe-based Postgrowth Innovation Lab, the surge in collective action for the right to repair flows through four disparate routes: consumer advocacy, environmental sustainability, communitarianism, and "creative tinkering and grassroots innovation." Those motivations attract a broad base of support. As of late 2023, thirty-three states and Puerto Rico in the United States had considered right-to-repair legislation, and Colorado, California, Minnesota, and New York had enacted laws.[271]

With degrowth, there will still be plenty of need for professional repairers, but everyone will have the right to fix stuff, whether it's their own or their friend's, independent of the market economy.

THE FIRST JOBS TO ABOLISH

Many jobs, in manufacturing, construction, maintenance, transportation, warehouse work, food production and processing, health care, sanitation, and other areas can be physically and mentally demanding, stressful, unpleasant, low status, poorly paid, or all of the above. But much of that work is essential to keep communities and countries functioning. Many such jobs would remain essential with degrowth, though in fewer numbers. They also would take different forms, with less exploitation and environmental impact. And currently undervalued jobs would be recognized with higher status and living wages. Working hours in such tough but essential occupations should also be distributed more broadly and equitably.

That said, it doesn't follow that all physically demanding, stressful, unpleasant occupations are essential. Some, like coal mining, have played important roles in growth economies, but they are devastating to our common future. A degrowing society will also have less use for, say, bomb makers and prison guards.

I suggest that in the United States, a heavy concentration of jobs that are both demanding and destructive can be found in the military-industrial complex. Almost three million people work just on the military side of the complex; they include active duty, reserve, and National Guard troops and civilian employees in the Defense Department. A healthy and degrowing United States will not need to justify keeping that many people employed in support of the resource-heavy, polluting, deadly work of war fighting, imperialism, and suppression of domestic dissent. Nor can we afford to keep almost a half million additional people employed by private military contractors.[272]

Dismantling the US war machine will bring great relief to hundreds of millions of people around the world, both within and outside US borders. Look at what our military has been up to in just the past six decades, with warfighting in Indochina, Grenada, Panama, Iraq, Kuwait, the Balkans, Afghanistan, Syria, Yemen, Iran, and more. Few now view any of those interventions as having been necessary or leading to positive outcomes. Furthermore, the US and the Soviet Union held the world hostage to nuclear holocaust throughout the forty-five-year Cold War, and the nuclear specter continues to haunt the world, with China, Pakistan, India, Israel, North Korea, France, and the United Kingdom all now possessing these weapons of mass destruction.

Negative global impacts have gone well beyond battlefield combat; the US government and corporations have long provided military support to repressive, murderous regimes around the globe, without any American boots hitting the ground. This reached peak depravity with the lavish, decades-long provision of weapons and support to Israel's settler-colonial government, which has committed exceptionally monstrous crimes against humanity in carrying out its genocide of Palestinians in Gaza and the West Bank.

Along with engaging in and enabling armed conflict, the Pentagon continuously damages everyday quality of life in eight-hundred-plus locations scattered across eighty countries where it has bases and other military installations. David Vine, the author of *Base Nation: How U.S. Military Bases Abroad Harm America and the World*, has summed up the pernicious impact of these sites:[273]

> While logic might seem to suggest that these bases make us safer, I've come to the opposite conclusion: in a range of ways our overseas bases have made us all less secure, harming everyone from US military personnel and their families to locals living near the bases to those of us whose taxes pay for the way our government garrisons the globe.[274]

Most of these locations were acquired by the US military for living and working spaces—which together include more than 170 golf courses, Vine notes—by kicking local people off their land. Dismantling the US war machine will be of incalculable benefit to people of the Global South, not only by closing overseas bases and removing direct threats of death and destruction but also by cutting off a major source of the greenhouse gases that threaten to destroy everyone's future in both the North and South.

The US military is the largest institutional user of petroleum-based fuels in the world and emits greenhouse gases with an annual global warming impact of sixty million metric tons of carbon dioxide.[275] If the Pentagon were a country, its emissions would rank just behind those of Ireland and Finland. They would surpass the annual emissions of Bulgaria, Croatia, and Slovenia combined. A large share of the US military's climate impact results from the construction, maintenance, and use of those eight hundred overseas properties they occupy, while most of the carbon emitted by their military operations comes from the combustion of jet fuel.[276]

The island of Guam in the Western Pacific was colonized by the United States in 1898. Today, almost one-third of its land area is occupied by the US military, and that occupation has long been ecologically destructive to the homeland of Guam's indigenous Chamoru people. In 2024, the public health writer Sara Mar reported that the situation was worsening dramatically with the establishment of a new US Marine base and firing range that together are sealing off and despoiling about five thousand acres of Chamoru land. The base area encompasses at least twenty officially recognized archeological sites and sits atop Guam's essential water source, an aquifer. The marine corps announced plans to fire more than five million rounds of ammunition on the range each year, creating a big, toxic lead-pollution hazard. Meanwhile, the air force sought a permit to burn or detonate thirty-five thousand pounds of old bombs and other munitions, mostly of World War II vintage, in Guam, essentially waging war on yet another stretch of Chamoru land. This, too, will create a severe health hazard and threaten the island's water supply. Mar quoted a Chamoru organizer who said the detonation and burning site was "close to everything. Migratory birds, sea turtles, fisheries . . . traditional medicines that grow in that area . . . and our water of course." [277]

Those are some of the reasons that elimination of almost three million war-related jobs in a degrowing United States would be welcomed by most of the world's people. But deep shrinkage of our military-industrial complex would also be a resoundingly positive development for people living within the United States—even, or maybe especially, for those currently employed by the military. It would also free up hundreds of billions of dollars every year to be spent instead on curbing global warming and fulfilling other urgent needs such as food and medical security; affordable, efficient housing; better, more affordable public transportation; and economic justice. That would include millions of socially valuable jobs, ones far more satisfying than employment in war making.

For enlisted personnel, having a job in the armed forces means being controlled and exploited. In 2023, enlisted troops in the four lowest ranks were being paid approximately $24,000 to $36,000 per year, before tax. Furthermore, reports *The Washington Post*, "Enlisted recruits often come disproportionately from poor communities."[278] A RAND Corporation report found that one in four military families suffers from food insecurity. Among families of junior enlisted troops, the rate jumps to an even more shocking 45 percent.[279] Spokane's *Spokesman-Review* elaborates:

> The RAND report found that more than 286,000 active-duty families experience "low food security," which the US Department of Agriculture defines as not getting an adequate quality, variety or amount of food. Of those, about 120,000 face "very low food security," reporting that they sometimes skip meals, eat less than they need or lose weight because they can't afford enough food.[280]

Despite being eligible for federal food programs, only 14 percent of those food-insecure military households had been receiving support from SNAP (formerly known as the food stamp program), the federal Women, Infants and Children (WIC) program, or even charitable food banks.[281]

A strategic retreat from the twenty-seven million acres of US territory now occupied by the military would end decades of land and water abuse.[282] Marine corps base camp Lejeune occupies about 160,000 acres on the North Carolina coast. The camp's potable water supply was long contaminated with gasoline and highly toxic chemicals, including trichloroethylene (TCE), perchloroethylene (PCE), and benzene. It has been called "the worst example of water contamination this country has ever seen."[283] TCE is associated with kidney, liver, cervical, and lymphatic cancers. PCE can cause leukemia, skin, colon, and lung cancer. Over a thirty-five-year period, more than a million troops and civilians drank, cooked with, or bathed in the camp's contaminated water.

Camp Lejeune remains up and running, but long-shuttered military bases continue to threaten Americans with their toxic legacies—what might be called zombie pollution. Consider Salina, Kansas, the city where I live. We face our own TCE threat, one created by Schilling Air Force Base, a Strategic Air Command station for nuclear-armed bombers that was closed down in 1965, after which the land and facilities were converted to civilian uses. By the 1990s, it became clear that the air force had left behind in the groundwater large amounts of TCE, once used as a degreaser by aircraft-maintenance crews. Further investigation revealed that an underground "plume" containing TCE and other toxic chemicals was slowly seeping eastward toward the aquifer that supplies Salina's water.[284]

Predictions of how long the plume would take to reach and ruin the aquifer if it's not stopped have varied from a decade to a century. A project to finally rein in the plume and decontaminate the TCE-laden soil between the former base and the water supply was approved in 2021. The work will take many years and cost $75 to $100 million. Officials express confidence that the plume can be stopped before it hits Salina's aquifer, but I haven't seen any promises in writing.[285]

After many years of delay, EPA placed a nationwide ban on TCE, to take effect in January 2025. But a regulatory freeze ordered that month by the incoming Trump administration put the ban on hold while officials decided whether to rescind it. The compound remains in use.[286]

People employed at military bases and those living in communities near bases face problems that extend beyond environmental damage. A nationwide statistical analysis showed that towns and cities abutting military installations had higher crime rates than towns that are not near bases but are similar in other respects. On average, property-crime rates were 19 percent higher in military than in civilian towns, while violent crime rates were 34 percent higher. These disparities were independent of any differences between the military- and civilian-adjacent towns in racial, ethnic, and gender composition, geographical region, and socio-economic conditions.[287]

Military bases and towns have also become notorious for high rates of sexual assault. That led the Defense Department to establish a Sexual Assault Prevention and Response Office (SAPRO), "the central authority charged with preventing sexual assault in the military and facilitating recovery for survivors," in the early 2000s.[288] The office publishes the total numbers of sexual assaults officially *reported* each fiscal year; in some years, it also estimates the *total* number of sexual assaults suffered by service members and civilian employees. The total number in 2021 was believed to be about forty-five thousand, victimizing 8.4 percent of all women in the armed forces and 1.5 percent of men. This estimated total was 13 percent lower than it had been a decade earlier, but at the same time, the number of victims reporting an assault increased by almost 150 percent. SAPRO's encouragement of service members to report assaults was clearly successful; while victims reported only one out of fifteen assaults in 2012, they reported one out of every five in 2021. Yet the biggest problem—those tens of thousands of sexual assaults every year—remains.[289]

The military's sexual assault scourge is not confined to the fifty states. Vine writes, "An estimated 30 percent of servicewomen are victimized during their time in the military and a disproportionate number of these crimes happen at bases abroad . . . In Okinawa and elsewhere, US troops have repeatedly committed horrific acts of rape against local women."[290] This grim era will come to an end only with the end of militarism.

SPREAD THE WORK, SPREAD THE SUFFICIENCY

The bullshit-job stories told to David Graeber by Betsy, Hannibal, Judy, and his other respondents are remarkable for their weirdness.

Such jobs, along with others that are simply cruel or destructive in their effects on the world—with war making perhaps at the top of that list—clearly must not accompany us well into the future. But what about countless other kinds of paid work that would not be needed in a degrowth economy, even if many in the current capitalist economy regard them as worthwhile and those holding the jobs find them rewarding?

I don't think anyone could plan with any confidence the employment mosaic of a degrowing United States. I realize that I've been passing a lot of judgment in this chapter, but I'm not a decision-maker. No one could or should be empowered to dictate a list of occupations scheduled for retention or elimination. And advocating for a twenty-hour workweek doesn't mean calling on some person or agency to decide which kinds of existing full-time jobs are socially necessary, reduce each to twenty hours per week, double the number of such jobs, and eliminate all others. No one should have the power to decree that one currently unpaid activity will become a paid job or that a certain paid job must henceforth be carried out by volunteers. How people spend their productive time in a degrowth future will depend on how the society in question reduces its input of energy and material resources and how it allocates the diminished quantities of resources to what are deemed essential functions—and how successful regular people are in wresting power from the economic elites who hold that power today.

Not long ago, many of the jobs people now hate didn't even exist, so we clearly can do without many of those undesirable work hours in the future. Fortunately, the very transformation into economies and societies that don't expend labor time on violating ecological limits will rid themselves of many other ills. On-the-job hours will be far outnumbered by hours lived away from work. More time will be left for food growing and other household self provisioning, caring for family, and joining in community efforts of all kinds. Unfair allocation of paid and unpaid work time will be gone, with far more equitable distribution both between the sexes and among racial and ethnic groups, and with job sharing playing an important role. Worker-run cooperatives will crowd out for-profit enterprises and the bullshit jobs they create.

Temporarily setting aside his usual reluctance, as an anarchist, to advocate for government policies, David Graeber argued that a society whose members spend, say, half as much time in paid work as they do today would require a guaranteed income for all. In that, he easily debunked one common trope: that with an assured, adequate income, people won't work. And he took head-on a common objection that he characterized neatly: "Most will work, but many will choose work that's of interest only to themselves. The streets would fill up with bad poets, annoying street mimes, and promoters of crank scientific theories, and nothing would get done."[291] Graber said not to worry about that prospect:

> No doubt a certain proportion of the population of a free society would spend their lives on projects most others would consider to be silly or pointless; but it's hard to imagine how it would go much over 10 or 20 percent. But already right now, 37 to 40 percent of workers in rich countries *already* feel their jobs are pointless . . . If we let everyone decide for themselves how they were best fit to benefit humanity, with no restrictions at all, *how could they possibly end up with a distribution of labor more inefficient than the one we already have?*[292]

I concur with those in the degrowth movement who advocate for a universal basic or living income and a workweek of no more than twenty hours. I also agree with those who are calling for universal basic services. Or, as others have suggested, universal income and universal services working in tandem could cover all such basic needs.[293]

A society that jettisons the profit motive and reduces exploitation of resources, but with fairness and sufficiency for all, will free the privileged from material excess, free the nonprivileged from scarcity and exploitation, and free everyone from exploitative and destructive jobs.

VI.

An Anthropause
of Our Own

Imagining the future world that will result from a phaseout of fossil fuels and well-thought-out, equitable degrowth, I've tried to catalog some of the serious problems that would be eliminated in the process. In that, I've focused on just a few snapshots from degrowth's expansive utopian vision.

Another vision, that of neoliberalism, has had the world in its grip for decades. But the neoliberal utopia, writes Giorgos Kallis, is monolithic and closed, an inalterable prescription. Degrowth utopias are the opposite, he argues. They're "not meant to provide a clear idea of where we should go; instead . . . they should open up our minds to a horizon of possibilities, which in turn shape our actions today."[294] As I was working on this book, I came across an essay that Timothée Parrique wrote for Céline Keller's book *Who's Afraid of Degrowth?* In the essay, Parrique lays out his own series of utopian snapshots. They're not happy subtractions like those in this book but rather happy additions out there on the horizon, waiting to be realized. You can read the full essay in Keller's book, but here are a few passages:

> Direct democracy in cooperatives with one person, one vote; participatory democracy in neighborhoods and municipal-

ities; and representative democracy at the bioregional and national level.

Sufficiency for all, excess for none.[295] Everyone is guaranteed free access to a decent level of food, education, health care, transportation, housing, water, energy, either via public services or by the granting of a basic income.

The pavement bursting open with flowers, fruit trees, and all the life that comes with it.

Imagine car-free garden cities where kids ride their bikes to school together, organized around the fifteen-minute rule. You can access all your daily activities by foot, bike, or public transport. Pedicars everywhere, alongside shared electric vehicles, free buses and trams, birdsongs in the streets, clean air, and swimmable rivers.[296]

In 2022, eight academics in environmental sciences and ecological economics published a *Nature* commentary outlining research questions that they feel must be answered when developing degrowth policies for the affluent Global North. They highlighted policies to deeply shrink destructive economic sectors—especially fossil fuel extraction, factory-farmed meat, motor vehicles, and aviation—while ensuring universal access to high-quality health care, food, transportation, housing, and renewable energy.[297] Most of their policy recommendations would work within existing political and economic systems. For example, they urged that companies "prioritize social and environmental benefits and take ecological costs into account" and that governments stop subsidizing fossil-fuel extraction and tax those other destructive industries. They recommended the development of decent living standards for all, along with reduced consumption of energy and material resources. They suggested learning from social forma-tions such as cooperatives, cohousing projects, and "transition towns." Somewhat more far-reaching was their declaration that housing should be treated as a fundamental need and right, not a profitable investment.

Elsewhere, these and other authors have written and spoken about the path to a degrowth society in much more radical terms. But their *Nature* commentary was about as heterodox as is allowed when one is writing for one of the world's premier sci-entific journals. The authors acknowledged that to achieve even these modest policies would require overcoming strong resistance from corporations, think tanks, lobbyists, and a large majority of Western politicians. Given those realities, elite resistance to even more far-reaching policies, such as a twenty-hour workweek or a fast-declining, leakproof cap on fossil fuel extraction and use, would likely be several times as ferocious.

In the United States, even if more farsighted lawmakers and policymakers were to gain majority control of the White House and both chambers in Congress, they would face the daunting task of building public support for policies unfamiliar and scary to most of the voters who had put them in office. They might overcome that resistance by leading with pitches that have the best chance of appealing to voters. Designating the oil, gas, and coal industries as public enemy number one would be a good start; large majorities dislike those companies. Leaders could then promise tight price controls on gasoline, diesel, and heating fuels—a sure winner—and follow that up with the pro-people, pro-equality, anti-capitalist economic policies that are staples of the degrowth vision.

It's hard to imagine such enlightened leaders coming to power anytime soon in this or any other economically dominant, cli-mate-clobbering countries. Degrowth advocates, therefore, have been wise to focus mostly on bottom-up social movements that start locally but aim to propagate widely. For inspiring precedents, degrowth advocates often turn to real-world examples: Indigenous societies who have never pursued economic growth, the Black-led environmental-justice movement,[298] worker-owned cooperatives, mutual-aid movements, citizens' assemblies, participatory municipal budgeting, and others. They cite the city of Barcelona, Spain, and the worldwide Fearless Cities movement it has inspired; the Mondragon Corporation in Spain's Basque region; Croatia's capital Zagreb; the global *buen vivir* movement that originated with the Indigenous struggle against oil extraction in Ecuador; the rural, women-led Via Campesina movement of two hundred million people around the world; and the Zapatista revolution in the Mexican state of Chiapas.

Kohei Saito discusses these and other models in his book *Slow Down: The Degrowth Manifesto*, which was published in English translation in 2024. In arguing for a future that he calls "degrowth communism," Saito refers to something very different from the production-obsessed, capital-C Communism of the Soviet Union and China. He's talking about a degrowth process that restores to our collective care of those "forms of wealth that should be managed and shared by every member of society"—especially "things like water, electricity, shelter, healthcare, and education," all of which should be treated as public goods and managed democratically. Crucially, he adds, "The means of production must be returned to the commons as well."[299]

Saito bases these and other conclusions on his reading of Karl Marx's unpublished writings from his final years, when Marx was deep into the study of ecology and Indigenous or traditional steady-state societies. He outlines the trajectory the younger Marx of the 1840s and 1850s (a period exemplified by *The Communist Manifesto*), who was all for growth and didn't consider ecological sustainability, to the Marx of the 1860s (when he wrote *Capital, Volume 1*), who was still a pro-growth productivist but had also become an eco-socialist focusing on sustainability, and, finally, to the Marx of the 1870s and 1880s. By then, writes Saito, Marx had turned against economic growth, because he saw, correctly, that it was incompatible with ecological sustainability. In Saito's view, Marx had become a "degrowth communist," though of course he never used that term himself.

In contrast to Marx and Engels' vision of lowercase-c communism, in which the state withers away, Saito writes, "The reform of democracy is more important now than ever before. This is because the power of the state will play an indispensable role in any effective response to climate change."[300] I think it's safe to say that most degrowth theorists agree that governments will always have a role to play, but that democracy—not only representative democracy but many other forms, including direct, inclusive, participatory, and deliberative democracy—is essential at all scales from the community to the nation-state.[301]

THE RESTING BRAKE

Degrowth is unmistakenly revolutionary. The twentieth-century philosopher Walter Benjamin, noting that Marx regarded revolutions as "locomotives of world history," wrote that a better metaphor might be "an attempt by the passengers on the train [. . .] to activate the emergency brake." Citing both Benjamin and Saito, the ecological organizer and degrowth advocate Andrew Ahern wrote in 2023:

> There is no shortage of things we could pull the emergency brake on that would allow us to halt ecological destruction with more speed than current capitalist interventions. As Saito and other degrowth thinkers have made clear, that might include the military, SUVs, private jets and car-cen-

tricity, industrial animal products, fast fashion, planned obsolescence, fossil fuels, single-use plastics, and more. Degrowth in these sectors would allow us to reduce our use of energy and materials and concentrate our focus on parts of the economy that provide well-being for people and the planet: renewable energy, plant-based foods, healthcare and education, democratic participation, and free time for friends, family, and nature. In this sense, Saito goes beyond Benjamin: more than the emergency brake, degrowth is a collective *resting* brake, a rest from capitalism, in which we finally slow down our impacts on the earth and each other. The care economy is a model for this kind of rest, as Saito puts it: "The more society shifts towards essential work that produces basic use-value, the slower the entire economy is likely to become." A combination of care, rest, and ecology is the future.[302]

I agree with Ahern. But there is a long, hard path from today's growth economy to one that has been freed from catastrophic growth and inequality, commodification of our lives and time, systemic racism, imperialism, and ecological cataclysms.

The social, economic, and political counterforces that obstruct our path to degrowth in the United States are formidable: an all-powerful private sector that depends on unimpeded growth for its very existence; dominant political parties fully invested in capitalist exploitation; a military-industrial complex that always gets its way; an electorate that largely accepts the myth that economic growth is essential; and, now, a Silicon Valley–centered billionaire establishment that claims it can remove all biophysical constraints on human domination of the earth and the universe beyond.

Each of those forces is driven by an insistence that neither the reversal of economic growth nor the treatment of the material world as a commons is possible. In arguing against such futility,

Giorgos Kallis, citing Serge Latouche, describes degrowth as a "concrete" utopia; that is, it has a "solid scientific basis" and is physically "doable." Anthropologists have thoroughly documented past and present societies that exemplify the degrowth vision. Economists have shown that a steady-state economy can be viable over the long haul *and* function better than a growth economy. Degrowth is not precluded by the laws of physics or the limits of biology. And the ossified classes and institutions that preserve and defend the growth economy are becoming increasingly fragile.[303]

The venerable World Social Forum's official slogan asserted that "another world is possible."[304] Whether to wallow in the dangers and miseries of our increasingly unjust and unsustainable economic system or work together to transform it is up to us.

Unfortunately, there's a grim mismatch between the extraordinarily compressed timeline available for preventing ecosystem collapse and the protracted political struggle required to achieve degrowth. The United States has been stuck for years in a bumper-to-bumper traffic jam of economic and social injustices, political threats, state-sponsored violence (both within and beyond our nation), public health catastrophes, and other crises. Immediate emergencies receive a prompt response, sometimes good, sometimes terrible, while "long emergencies" like climate change are met only with procrastination. The need for thoroughgoing ecological transformation—a perennial back burner issue—has been knocked completely off the stovetop by this accelerating onslaught of crises.

Even if a rapid phaseout of fossil fuels, reduced production, transformation of work, radical equality, and other issues can achieve some breakthroughs, the rich and powerful are dug in and ever willing to use violence. Notably, they and their supporters have most of America's state and civilian firepower on their side.

Kallis offers this sobering prospect: "If history is any guide, the material and political changes involved in a degrowth scenario are unlikely to be easy. The end of fossil fuels will increase pressure for redistribution. We know that substantial redistribution seldom takes place without war or great destruction."[305] This is self-evident in the case of a government like the one in Washington, with its decades-old, bipartisan policy of violently attacking people's movements in this country and around the world. A government that has supported state violence on the streets and in prisons within its borders, maintained a massive armed presence abroad, and bankrolls genocide in Palestine won't hesitate to use force in defense of its ecocidal economy.

The power of the state and capitalism, however, is no reason to set aside the struggles against fossil fuels and ecological destruction. There's clearly no guarantee of success in those struggles, but there is nothing about the goals themselves that makes them inherently unachievable. Many societies throughout history and prehistory have lived largely in harmony with nature, equitably, with low energy and material inputs, and everyone can learn from how they integrated into ecosystems rather than dominating them. But if those of us in industrialized societies are to find a way to follow their example, much will still be left for us to figure out on our own. Those communities who've had the wisdom to live without abusing nature, in ways that can be sustained for the long haul, needed only to continue doing so. In contrast, nations like the United States could never have become the global behemoths they are today without gorging on the fossil-fuel bonanza of the past century and a half. Those unprecedented quantities of energy enabled us to create a built environment, transportation systems, and an entire society that is wholly dependent on continued resource extraction and a far greater energy input than the sun can provide. We're now faced with the necessity of doing something unprecedented: voluntarily backing out of our high-energy, high-resource-use cul-de-sac and adapting to a materially modest way of life. We'll be far better off if we can manage to do so, but overcoming resistance from above will be far from easy.

If the degrowth movement were betting solely on the success of national and international political decisions, the project would be a grim endeavor indeed. But that's not the case; the movement is placing most of its chips on local community- and workplace-based transformation. With global action to head off climate chaos and ecological catastrophe having run for decades at the speed of molasses in January (and recently, it seems, even running in reverse), it's even more important to take the degrowth road now, collectively, where we live. When a thousand local degrowth flowers bloom, the material and ecological foundations of civilization will crumble more slowly, giving societies more time to adapt; by the time a million bloom, we or those who follow us may finally live in a beautiful, just, nature-centered world that humanity should have been working toward all along. Communities or regions that get an early start on practicing material frugality, serving the collective good, and ensuring sufficient and equitable satisfaction of needs for all will be better prepared for a more resource-limited, less destructive civilization of the future.

AN ANTHROPAUSE OF OUR OWN

Researchers studying the songs and breeding behavior of the white-crowned sparrow in urban portions of the San Francisco Bay Area found that when human-generated noise levels (mostly from vehicle traffic) dropped sharply during the 2020 anthropause, the birds immediately went back to vocalizing as their ancestors did in the much quieter San Francisco of the 1950s. This allowed them to, in the authors' words, produce "higher performance songs" of lower volume, maximizing "communication distance and salience." Better songs generally mean greater breeding success, and sure enough, further studies found more successful mating among songbirds in general during the anthropause. I'd venture a guess that for just a few weeks, those were some very happy sparrow communities.[306]

The birds found their voice because humans temporarily set aside some of their mechanisms for dominating nature. Our efforts to sustain human communities are not at the root of the ecological crisis. It's dominant societies' for-profit degradation of nature and their voracious appetite for industrial energy and other resources that are to blame. When birds sang softly and clearly once again in San Francisco, and hosts of other creatures returned after long absences to human-dominated landscapes during the 2020 anthropause, it wasn't because humans had disappeared. We were still there; our fellow animals reverted to their "natural" behaviors and returned to their former habitats because we had temporarily stopped doing a lot of the stuff that makes those places inhospitable to their way of life (and to ours).

When we manage to stop abusing the living world—an each other—our urban, suburban, and rural areas will become far more hospitable, not only to other plant and animal species but to human life as well. We'll finally be free to enjoy a beautiful anthropause of our own.

Afterword

When I began writing this book in mid-2023, the prospects for rebuilding US society on principles of degrowth were as dim as ever. Economic growth remained the goal that eclipsed all others, no matter the human and ecological cost. The United States pumped more crude oil that year than any country in the world had ever pumped in one year.[307] Economic inequality was rising fast. Consequently, as I began to write a book envisioning a collective future in which we are liberated from capitalism, overproduction, and overconsumption, I knew I risked prompting skepticism among readers who might ask why I would write about such a vision for our country, given its record as a global leader in greenhouse gas emissions, material excess, militarism, bigotry, and the enforcement of inequality.

The outlook would soon turn even bleaker, as the Biden administration and both parties in Congress armed and supported Israel's genocide in Gaza through fifteen months of escalating savagery. Then when Donald Trump returned to power in 2025, he and Congress took up where Biden left off, with no end to the horror in sight.

As the months passed and Washington doubled and tripled down on Israel's campaign to eradicate Palestinian society, I anticipated that some readers of *Anthropause* might have plenty of questions for me. For example, can we ever trust a government that has expended billions of dollars per month on aiding and abetting a genocide to start working toward just, humane, ecologically sound degrowth? Will a White House and Congress that tacitly supported an aid blockade that starved countless Palestinian children to death—and then installed a food-aid program under which Israeli soldiers would shoot and kill people as they waited to receive the aid—decide to end their long-term neglect of the eighteen million US households that don't have enough to eat? Having enabled the destruction of more than half a million Palestinian homes, will they ever start recognizing a universal right to decent housing in this country? Given that they backed Israeli forces in the murder of dozens of doctors and targeted all of Gaza's hospitals for destruction, do you think they will ever stop denying Americans the right to universal health care? When they've countenanced the abduction, imprisonment, and torture of countless Palestinians for decades, will they ever end the imprisonment of one in five Black American men? And is it possible that our government, which for decades endorsed Israel's invasion, occupation, and creeping annexation of Palestinian homelands, would ever consider returning large swaths of federal land to the Native tribes from whom they stole them? Or close the eight hundred US-military bases that are squatting on other people's land around the world?

I recently heard an interview with Yanis Varoufakis, the secretary-general of the Democracy in Europe Movement and a former Greek finance minister, in which he laid out with passionate succinctness the damage that the West's toleration of Israel's crimes has wreaked:

> The Palestinian genocide is the great moral clarifier of our times. Anyone who turns a blind eye to what is happening in Palestine today . . . or is frightened of being denounced

if they don't denounce pro-Palestine demonstrations, for instance—they lose their soul. I don't mean this in a spiritual or religious sense; I am an atheist. What I mean is that it becomes very hard for them to resist any abuse of power when it comes to the rights of women, the rights of workers, when it comes to their own municipalities' right to maintain decent services. You see, it cascades. Once you turn a blind eye to Palestine, then your capacity for ethically resisting any wrong, any injustice that is done in your own society, your own region, your own country simply dissipates . . . This is how fascism takes hold of a society. One morally clarifying issue is enough to start a domino effect, and the result is a society that cannot be defended by its own members when the fascists come knocking on the door.[308]

Meanwhile, another global tragedy, the climate emergency, continued to go unaddressed. Not even bothering to feign concern over global warming as the Democrats had done, the Republicans actively promoted the burning of fossil fuels, acceleration of resource extraction, abuse of nature, and the abandonment of environmental regulation. Visions of justice and sufficiency for all, excess for none, a flourishing of nature, and a better quality of life for everyone appeared to be going up in smoke.

Certainly, the future I was envisioning in *Anthropause* was becoming even less likely to emerge. However, I realized that such realities didn't undermine my core argument. This was never a book of prophecy. It's more like a Hail Mary pass; the odds against success are rising, but we still have the ball. Ahead of us are no certainties but some appealing possibilities, all of them contingent on the emergence of a mass movement rooted in degrowth principles. When I started writing, a radical groundswell of that sort was already improbable, especially in the United States and other affluent Western nations, and it remains so. However, to predict that achieving degrowth will be difficult is not to predict that people in this country or around the world will just sit on their hands and let the worst come to pass.

One way or another, economic growth will eventually cease worldwide and then shift into reverse. The nature of that shrinkage will depend on what people do, collectively, in the next decade or two. If we continue barreling ahead with business as usual—overproduction for profit, exploitation of people and nature, neglect of human needs, more militarism, more imperialism, more genocide—the reversal of growth will come in the form of chaotic collapse. However, with careful, equitable, intentional degrowth, a more modest but livable future remains possible. Whatever our prospects, Western societies must act as if the world *can* have a livable future. What we should not do, however, is to expect that a better future will look anything like our present.

Notes

CHAPTER I: THE DAYS THE EARTH STOOD STILL

1. Zhu Liu et al., "Near-Real-Time Monitoring of Global CO_2 Emissions Reveals the Effects of the COVID-19 Pandemic," *Nature Communications* 11 (2020): 5172, https://doi.org/10.1038/s41467-020-18922-7; Komathi Kolandai et al., "Anthropause Appreciation, Biophilia, and Ecophilosophical Contemplations Amidst a Global Pandemic," *Journal of Environmental Psychology* 85 (2023): 101943, https://www.sciencedirect.com/science/article/pii/S0272494422001888.

2. "The Urban Wild: Animals Take to the Streets Amid Lockdown—in Pictures," *Guardian*, April 22, 2020, https://www.theguardian.com/world/gallery/2020/apr/22/animals-roaming-streets-coronavirus-lockdown-photos; Emma Gatten, "World Without Humans: Coronavirus Lockdown Triggers Giant Experiment in How Animals Live Without Us," *Telegraph*, April 12, 2020, https://www.telegraph.co.uk/news/2020/04/12/scientists-using-coronavirus-lockdown-understand-wildlife-behaves/.

3. The term "anthropause" was first used by Christian Rutz et al., "COVID-19 Lockdown Allows Researchers to Quantify the Effects of Human Activity on Wildlife," *Nature Ecology & Evolution*, no. 4 (2020): 1156–59. Some subsequent authors have capitalized the word anthropause. I am using Rutz et al.'s original lower case.

4. Lynn Williams et al., "What Have We Learned About Positive Changes Experienced During Covid-19 Lockdown? Evidence of the Social Patterning of Change," *PloS One* 16, no. 1 (2021): e0244873, https://journals.plos.org/plosone/article?id=10.1371/journal.pone.0244873.

5. Patrick Van Kessel et al., "In Their Own Words, Americans Describe the Struggles and Silver Linings of the COVID-19 Pandemic," Pew Research Center, March 5, 2021, pewresearch.org/data-labs/2021/03/05/in-their-own-words-americans-describe-the-struggles-and-silver-linings-of-the-covid-19-pandemic/.

6. Herb Scribner, "'Indoor Generation': Here's How Much Time We Spend Indoors," *Deseret News*, May 16, 2018, https://www.deseret.com/2018/5/17/20645140/indoor-generation-here-s-how-much-time-we-spend-indoors/.

7. Dehui Geng et al., "Impacts of COVID-19 Pandemic on Urban Park Visitation: A Global Analysis," *Journal of Forestry Research* 32, no.2, (2021): 553–67. City parks of course remained closed where there were blanket lockdowns in place.

8. Edward O. Wilson, "Biophilia and the Conservation Ethic," in *Evolutionary Perspectives on Environmental Problems*, ed. D. J. Penn and I. Mysterud (Routledge, 2017), 249–58.

9. Kolandai et al., "Anthropause."

10. Nathan Young et al., "Is the Anthropause a Useful Symbol and Metaphor for Raising Environmental Awareness and Promoting Reform?" *Environmental Conservation* 48, no. 4 (2021): 274–77.

11. Adam Searle et al., "After the Anthropause: Lockdown Lessons for More-than-Human Geographies," *The Geographical Journal* 187, no. 1 (2021): 69–77. Recent years have seen proposals for a permanent anthropause of sorts. Perhaps most prominent has been the "Half Earth" concept, under which lands and seas adding up to about 50 percent of the earth's surface area would be conserved in their "natural" state. The name comes from the title of the late eminent biologist E.O. Wilson's 2016 book *Half-Earth: Our Planet's Fight for Life*; it has also emerged in the form of a movement called "Nature Needs Half" and has won much media attention and increasing support in societies of the Global North. The 50-percent proposition has caught flak on several counts. In some of its various forms, it calls for physical separation of humans from the bulk of "unspoiled" nonhuman nature (which would be a tragedy for our species) and would intensify industrialization and its impacts in the half of the planet that's left unprotected. It would cut many traditional farming, pastoralist, and hunting-gathering communities off from their means of subsistence. To some, it seemed akin to the "ecomodernist" vision of a hyper-technological society that can supply all human necessities to those who can afford them while interacting almost not at all with the earth's ecosystems. Elsewhere, it was regarded as a slap to the face of Indigenous cultures, who have always known how to live in harmony with the natural world. Edward O. Wilson, *Half-Earth: Our Planet's Fight for Life* (WW Norton & Company, 2016); Harvey Locke, "Nature Needs Half: A Necessary and Hopeful New Agenda for Protected Areas," *Nature New South Wales* 58, no. 2 (2014): 7–17; Harvey Locke, "Nature Needs (At Least) Half: A Necessary New Agenda for Protected Areas," in *Protecting the Wild: Parks and Wilderness, the Foundation for Conservation*, ed. G. Wuerthner, E. Crist, and T. Butler (Island Press, 2015), 3–15; Jeremy Hance, "Scientists Call for a Paris-Style Agreement to Save Life on Earth," *Guardian*, June 28, 2018, https://amp.theguardian.com/environment/radical-conservation/2018/jun/28/scientists-call-for-a-paris-style-agreement-to-save-life-on-earth; Austin Miles, "Colonial Ecologies of the Half Earth," *Resilience*, April 6, 2022, https://www.resilience.org/stories/2022-04-06/colonial-ecologies-of-the-half-earth/.

12. Ekaterina Chertkovskaya, "Degrowth," in *Handbook of Critical Environmental Politics* (Edward Elgar, 2022), 116–28; Andrew Ahern, "Red and Green Make . . . Degrowth: On Kohei Saito's 'Marx in the Anthropocene'," *Los Angeles Review of Books*, July 23, 2023, lareviewofbooks.org/article/red-and-green-make-degrowth-on-kohei-saitos-marx-in-the-anthropocene/.

13. degrowthjournal.org.

14. Giorgos Kallis, *Degrowth* (Agenda, 2018), 9.

15. Discussion and definition of the *commons* was distorted for decades thanks to a viral article titled "The Tragedy of the Commons" by Garrett Hardin, published in the journal Science in 1968. In it, Hardin described a *commons* as a finite resource that is subject to ungoverned exploitation. (He famously used an unfenced pasture to illustrate his point.) In a fiftieth-anniversary look back at the article, Brett Frischmann et al. wrote, "Hardin confused resources with governance. In his sheepherding allegory, for example, the relevant resource is a pasture, and the relevant governance is open-access sharing: as the allegory begins, 'Picture a pasture open to all.' To describe commons as the resource subject to tragedy is a category error. Commons are *not*, and should not be conflated with, resources. They are neither common-pool resources nor public goods; these types of sharable goods may, however, be governed as or within commons. Instead, commons are a form of resource governance where members of a community share resources on terms set by the community," Garrett Hardin, "The Tragedy of the Commons," Science 162 (1968) : 1243–48, https://www.taylorfrancis.com/chapters/edit/10.4324/9781315695730-62/tragedy-commons-garrett-hardin; Brett Frischmann et al., "Retrospectives: Tragedy of the Commons After 50 Years," *Journal of Economic Perspectives* 33, no. 4 (2019): 211–28.

16. Kallis, *Degrowth*, 118–22.

17. Ivan Savin and Lewis King, "Idea of Green Growth Losing Traction Among Climate Policy Researchers, Survey of Nearly 800 Academics Reveals," *Conversation*, September 20, 2023, https://theconversation.com/idea-of-green-growth-losing-traction-among-climate-policy-researchers-survey-of-nearly-800-academics-reveals-213434.

18. Buzz around the "abundance agenda" peaked with publication of Ezra Kelin and Derek Thompson's bestseller titled (what else?) *Abundance* (Simon and Schuster, 2025). It's largely a rehash of the green growth and Green New Deal visions, the only novel element being a demand for sweeping deregulation of business to jump-start technological innovation and buildout of infrastructure. The biophysical and ecological evidence against green growth has been articulated widely and repeatedly (See for example, Jason Hickel, *Less Is More: How Degrowth Will Save the World* (Random House, 2020) and the many studies I cite in Stan Cox, *The Green New Deal and Beyond: Ending the Climate Emergency While We Still Can* (City Lights, 2020). Those environmental and resource issues are the most fundamental failings of *Abundance*, but Klein and Thompson have also come under well-founded attack for favoring business interests over civil society, ignoring or denying any difference between the interests of owners and workers, neglecting economic egalitarianism and the social safety net, and insisting on freeing business from regulation, which will inevitably weaken environmental and labor protections. Malcolm Harris, "What's the Matter with Abundance?" *The Baffler*, March 18, 2025, https://thebaffler.com/latest/whats-the-matter-with-abundance-harris; Matt Breunig, "What the 'Abundance Agenda' Leaves Out," *Jacobin*, March 24, 2025, https://jacobin.com/2025/03/abundance-klein-thompson-book-review; Nathan J. Robinson, "Abandon 'Abundance'", *Current Affairs*, June 13, 2025, https://www.currentaffairs.org/news/abandon-abundance.

19. Timothée Parrique, *The Political Economy of Degrowth* (PhD dissertation, Stockholms Universitet, 2019), theses.hal.science/tel-02499463/file/2019CLFAD003_PARRIQUE.pdf.

20. Parrique, "The Political."
21. Giorgos Kallis, "In Defense of Degrowth," *Ecological Economics* 70, no. 5 (2011), 873–80, cited by Parrique, *The Political*.
22. S. Latouche, *Vers une société d'abondance frugale : Contresens et controversies sur la décroissance.* (Mille et une Nuits, 2011), cited by Parrique, *The Political*, 225.
23. Parrique, *The Political*, 228.
24. Ibid, 229.
25. Céline Keller, *Who Is Afraid of Degrowth?* (Flyeralarm, 2020), ii; print copies or pdf download at celinekeller.com/who-is-afraid-of-degrowth.
26. Keller, 59–60. The Kallis comments quoted by Keller come from his 2021 video titled "What Is Degrowth?" Posted June 29, 2021, YouTube, 31 min., 21 sec., youtube.com/watch?v=alp2ZJnvwW8.
27. Larry Edwards and Stan Cox, "Cap and Adapt: Failsafe Policy for the Climate Emergency," *Solutions* 11 (2020): 22–31, https://landinstitute.org/scientific-pub/cap-and-adapt-failsafe-policy-for-the-climate-emergency/.
28. Brett Christophers, "We Are Taking a Devastating Risk with the Green Energy Sector—One That Might Cost Us Our Future," *Guardian*, February 27, 2024, https://www.theguardian.com/commentisfree/2024/feb/27/climate-crisis-private-sector-government-investment.
29. Milena Büchs, "Sustainable Welfare: How Do Universal Basic Income and Universal Basic Services Compare?" *Ecological Economics* 189 (2021): 107152, https://www.sciencedirect.com/science/article/pii/S092180092100210X.
30. Elhacham et al., "Global Human-Made Mass Exceeds All Living Biomass." *Nature* 588, no. 7838 (2020): 442–44. The total human-made mass does not include wastes.
31. United Nations, "Global Material Flows Database," resourcepanel.org/global-material-flows-database.
32. Needed changes in land use will be addressed in coming chapters.
33. US Energy Information Administration, "United States Produces More Crude Oil than Any Country, Ever," March 11, 2024, eia.gov/todayinenergy/detail.php?id=61545.
34. This research is summarized, with citations, in Stan Cox, *The Green New Deal and Beyond: Ending the Climate Emergency While We Still Can* (City Lights, 2020), 52–71.
35. Richard Wilkinson and Kate Pickett, "Why the World Cannot Afford the Rich," *Nature* 627, no.8003 (2024): 268–70.

CHAPTER II: TO ENSURE DOMESTIC TRANQUILITY

36. *2020 Census Urban Areas Facts* (US Census Bureau, 2023), census.gov/programs-surveys/geography/guidance/geo-areas/urban-rural/2020-ua-facts.html.
37. Richard Heinberg, "The Gasoline-Powered Leaf Blower as a Metaphor for Industrial Society," *Resilience*, November 1, 2023, resilience.org/stories/2023-11-01/the-gasoline-powered-leaf-blower-as-a-metaphor-for-industrial-society.
38. Heinberg, "Gasoline"; Jackie DiFrancesco, "Grounds for Change: Reducing Noise Exposure in Grounds Management Professionals," US Centers for Disease Control (blog), July 25, 2018; updated November 25, 2024, blogs.cdc.gov/niosh-science-blog/2018/07/25/landscape-noise1; Mark Nevitt, "Think Globally on Climate, Act Locally on Leaf Blowers," *Regulatory Review*, February 6, 2023, papers.ssrn.com/sol3/papers.cfm?abstract_id=4359140.

39. Erica Walker and Jamie Banks, "Characteristics of Lawn and Garden Equipment Sound: A Community Pilot Study," *Journal of Environmental and Toxicological Studies* 1, no. 1 (2017): 10-16966; Farhad Forouharmajd et al. "Is Enough Attention Paid to the Health Effects of LowFrequency Noise in Today's Society?" *International Journal of Preventive Medicine* 13 (2023): 162, https://pmc.ncbi.nlm.nih.gov/articles/PMC9999102.

40. Walker and Banks, "Characteristics," cited by Tony Dutzik et al., *Lawn Care Goes Electric*, PIRG, October 2023, publicinterestnetwork.org/wp-content/uploads/2023/10/Lawn_Care_Goes_Electric_Oct23.pdf?ssp=1&darkscheme-ovr=1&setlang=en-US&safesearch=moderate.

41. Walker and Banks, "Characteristics."

42. DiFrancesco, *Grounds*; Jo Anne Balanay et al., "Assessment of Occupational Noise Exposure Among Groundskeepers in North Carolina Public Universities," *Environmental Health Insights* 10 (2016): EHI-S39682, https://journals.sagepub.com/doi/full/10.4137/EHI.S39682.

43. "Provide Hearing Protection," National Institute for Occupational Safety and Health, February 16, 2024, cdc.gov/niosh/noise/prevent/ppe.html. Here's a sample of the verbiage: "Hearing protector fit testing generates a Personal Attenuation Rating (PAR), [which] estimates an individual worker's reduction in noise exposure when using that hearing protector. A worker's protected noise exposure may be determined by subtracting the PAR from the measured exposure level."

44. SA Alamin Gabasa et al., "Vibration Transmitted to the Hand by Backpack Blowers," *International Journal of Automotive and Mechanical Engineering* 16, no. 2 (2019): 6697–705.

45. See Stan Cox, "We Deserve a Quieter World," *Nation*, September 18, 2024, https://www.thenation.com/article/society/noise-pollution-mental-health: "With every AI project abandoned, every bitcoin not mined, every pickup truck not sold, every jet fighter not flown, people somewhere will get relief from noise pollution."

46. Bianca Bosker, "Why Everything Is Getting Louder," *Atlantic*, November, 2019, https://www.theatlantic.com/magazine/archive/2019/11/the-end-of-silence/598366/.

47. Bosker, "Getting Louder."

48. Allyson Chiu, "The Problem with Gas-Powered Leaf Blowers," *Washington Post*, November 5, 2023, https://www.washingtonpost.com/climate-solutions/2023/11/05/leaf-blowers-fall-environment-health.

49. Dutzik et al., "Lawn Care."

50. Allyson Chiu, "Why You Should Be Lazy and Leave Your Leaves in the Yard," *Washington Post*, October 14, 2023, https://www.washingtonpost.com/climate-solutions/2023/10/14/raking-leaves-yard-fall-environment.

51. Chiu, "Why You Should Be Lazy."

52. Ivan Illich, *La Convivialité* (Éditions du Seuil, 1973), quoted by Timothée Parrique, *The Political Economy of Degrowth* (PhD dissertation, Stockholms Universitet, 2019), theses.hal.science/tel-02499463/file/2019CLFAD003_PARRIQUE.pdf.

53. Fabio Falchi et al., "The New World Atlas of Artificial Night Sky Brightness," *Science Advances* 2, no. 6 (2016): e1600377; Zhiheng Chen et al., "Using Mobile Phone Big Data to Identify Inequity of Artificial Light at Night Exposure: A Case Study in Tokyo," *Cities* 128 (2022): 103803, https://www.sciencedirect.com/science/article/abs/pii/S0264275122002426.

54. Christopher Kyba et al., "Citizen Scientists Report Global Rapid Reductions in the Visibility of Stars from 2011 to 2022," *Science* 379, no. 6629 (2023): 265–68.

55. Kyba, "Citizen."

56. Ibid.

57. Terrel Gallaway, "On Light Pollution, Passive Pleasures, and the Instrumental Value of Beauty," *Journal of Economic Issues* 44, no. 1 (2010): 71–88.

58. Kyba, "Citizen."

59. Ibid.

60. Gallaway, "On Light."

61. Travis Longcore and Catherine Rich, "Ecological Light Pollution," *Frontiers in Ecology and the Environment* 2, no. 4 (2004): 191–98.

62. Amedeo Argentiero et al., "Outdoor Light Pollution and COVID-19: The Italian Case," *Environmental Impact Assessment Review* 90 (2021): 106602, https://www.sciencedirect.com/science/article/pii/S0195925521000524; Longcore and Rich, "Ecological."

63. Saioa Legarrea Imizcoz and María Ángeles Marcos García, "Fewer Insects Hitting Your Car Windscreen? Here's Why," *Conversation*, November 8, 2023, https://the-conversation.com/fewer-insects-hitting-your-car-windscreen-heres-why-216544.

64. K.M. Zielinska-Dabkowska et al., "Reducing Nighttime Light Exposure in the Urban Environment to Benefit Human Health and Society," *Science* 380, no.6650 (2023): 1130–35.

65. S.M. Pawson and MK-F. Bader, "LED Lighting Increases the Ecological Impact of Light Pollution Irrespective of Color Temperature," *Ecological Applications* 24, no.7 (2014): 1561–68.

66. Candace Fallon et al., "Evaluating Firefly Extinction Risk: Initial Red List Assessments for North America," *PLoS One* 16, no. 11 (2021): e0259379. They report that sufficient data for estimating extinction risk is available for 47 percent of North American firefly species and that of those species, 30 percent are at risk.

67. Jae-Jun Kim et al., "Biologically Inspired LED Lens from Cuticular Nanostructures of Firefly Lantern," *Proceedings of the National Academy of Sciences* 109, no. 46 (2012): 18674–78.

68. Pawson and Bader, "LED"; Gallaway, "On Light."

69. *Artificial Light at Night: State of the Science 2022* (International Dark-Sky Association, 2022), darksky.org/news/artificial-light-at-night-state-of-the-science-2022-report.

70. Jesús Rodrigo-Comino et al., "Light Pollution: A Review of the Scientific Literature," *The Anthropocene Review* 10, no. 2 (2023): 367–92.

71. Shawna Nadybal et al., "Light Pollution Inequities in the Continental United States: A Distributive Environmental Justice Analysis," *Environmental Research* 189 (2020): 109959, https://www.sciencedirect.com/science/article/pii/S0013935120308549; Zielinska-Dabkowska, et al., "Reducing"; Argentiero et al., "Outdoor Light"; Filippo Crea, "Light and Noise Pollution and Socioeconomic Status: The Risk Factors Individuals Cannot Change," *European Heart Journal* 42, no. 8 (2021): 801–804.

72. Nadybal, "Light Pollution."

73. Ibid.

74. Zielinska-Dabkowska et al., "Reducing"; Rebecca Steinbach et al., "The Effect of Reduced Street Lighting on Road Casualties and Crime in England and Wales:

Controlled Interrupted Time Series Analysis," *Journal of Epidemiology and Community Health* 69, no. 11 (2015): 1118–24.

75. Nadybal, "Light Pollution."

76. Stan Cox, *Losing Our Cool: Uncomfortable Truths About Our Air-Conditioned World* (The New Press, 2010); Stan Cox, "In the Heat Wave, the Case Against Air Conditioning," *Washington Post*, July 11, 2010, https://www.washingtonpost.com/archive/opinions/2010/07/11/try-to-imagine-dc-without-so-much-ac/18602c74-b345-4e2f-9407-9f2f5dc06f2b; Stan Cox, "With Air-Conditioning, Have We Passed the Point of No Return?" *Nation*, June 17, 2024, https://www.thenation.com/article/environment/air-conditioning-climate-change-emissions; Stan Cox, "I Swore Off Air Conditioning, and You Can Too," *New York Times*, August 31, 2024, https://www.nytimes.com/2024/08/31/opinion/heat-wave-air-conditioning-climate-change.html.

77. Yabin Dong et al., "Greenhouse Gas Emissions from Air Conditioning and Refrigeration Service Expansion in Developing Countries," *Annual Review of Environment and Resources* 46, no. 1 (2021): 59–83.

78. James McClung et al., "Exercise-Heat Acclimation in Humans Alters Baseline Levels and Ex Vivo Heat Inducibility of HSP72 and HSP90 in Peripheral Blood Mononuclear Cells," *American Journal of Physiology-Regulatory, Integrative and Comparative Physiology* 294, no. 1 (2008): R185–R191; R. De Dear et al., "A Review of Adaptive Thermal Comfort Research Since 1998," *Energy and Buildings* 214 (2020): 109893, https://www.sciencedirect.com/science/article/pii/S0378778819337910.

79. M.J. Mendell et al., "Elevated Symptom Prevalence Associated with Ventilation Type in Office Buildings," *Epidemiology* 7, no. 6 (1996), 583–89; M.J. Mendell and A.H. Smith, "Consistent Pattern of Elevated Symptoms in Air-Conditioned Office Buildings: A Reanalysis of Epidemiologic Studies," *American Journal of Public Health* 80, no. 10 (1990), 1193–99; G.S. Graudenz et al., "Association of Air-Conditioning with Respiratory Symptoms in Office Workers in Tropical Climate," *Indoor Air* 15, no. 1 (2005), 62–66; P. Preziosi et al., "Air-Conditioning at Workplace and Health Services Attendance in French Middle-Aged Women: A Prospective Cohort Study," *International Journal of Epidemiology* 33, no. 5 (2004), 1120–23.

80. Marina Romanello et al., "The 2021 Report of the Lancet Countdown on Health and Climate Change: Code Red for a Healthy Future," *The Lancet* 398, no. 10311 (2021): 1619–62.

81. Cox, *Losing*, 94.

82. Tiffany Hsu, "The Advertising Industry Has a Problem: People Hate Ads," *New York Times*, October 28, 2019, https://www.nytimes.com/2019/10/28/business/media/advertising-industry-research.html.

83. Hsu, "Advertising"; Peter Noel Murray, "Why Consumers No Longer Like Advertising," *Psychology Today*, July 23, 2020, https://www.psychologytoday.com/us/blog/inside-the-consumer-mind/202007/why-consumers-no-longer-advertising.

84. Kate Lindsay, "Something Went Terribly Wrong with Online Ads," *Atlantic*, February 27, 2024, https://www.theatlantic.com/technology/archive/2024/02/online-ads-more-annoying/677576/.

85. Jonathan Ross Gilbert et al., "The Dance Between Darkness and Light: A Systematic Review of Advertising's Role in Consumer Well-Being (1980–2020)," *International Journal of Advertising* 40, no. 4 (2021): 491–528.

86. Vance Packard, *The Hidden Persuaders* (Ig Publishing, 2007), 12–13.

87. Packard, *Hidden*.

88. Ross Gilbert et al., "Dance."

89. Cansu Oral and Joy-Yana Thurner, "The Impact of Anti-Consumption on Consumer Well-Being," *International Journal of Consumer Studies* 43, no. 3 (2019): 277–88; Ross Gilbert et al., "Dance."

90. Matthew McAllister, "But Wait, There's More!: Advertising, the Recession, and the Future of Commercial Culture," *Popular Communication* 8, no. 3 (2010): 189–93.

91. Patrick Hartmann et al., "Perspectives: Advertising and Climate Change–Part of the Problem or Part of the Solution?" *International Journal of Advertising* 42, no. 2 (2023): 430–57.

92. Robert Istrate et al., "The Environmental Sustainability of Digital Content Consumption," *Nature Communications* 15, no. 1 (2024): 3724.

93. I use "poor" and "low-wage" as defined by the Poor People's Campaign: Olivia Rosane, "'We Are a Resurrection': Poor People's Campaign Rallies for Low-Wage Voters in DC," *Common Dreams*, June 30, 2024, commondreams.org/news/poor-peoples-campaign-voters.

94. Peter Menzel, *Material World: A Global Family Portrait* (Counterpoint, 1994).

95. The context in which Veblen made this quip isn't entirely clear, but he is widely credited with it: goodreads.com/quotes/88689-invention-is-the-mother-of-necessity. Plato wrote in *Republic, Book II*, "The true creator is necessity who is the mother of invention."

96. Stan Cox, "Enough for Everyone" with Tracy Matsue Loeffelholz and Jana Frederick Sanders, "Energy, Housing, Food, Water: What's a Fair Share?" *Yes! Magazine*, Fall, 2021, 20–29, https://www.yesmagazine.org/issues/how-much-is-enough.

97. Mark Perry, "New US Homes Today Are 1,000 Square Feet Larger Than in 1973 and Living Space per Person Has Nearly Doubled," American Enterprise Institute, June 05, 2016, aei.org/publication/new-us-homes-today-are-1000-square-feet-larger-than-in-1973-and-living-space-per-person-has-nearly-doubled/.

98. Jan Christoph Steckel et al., "Development Without Energy? Assessing Future Scenarios of Energy Consumption in Developing Countries," *Ecological Economics* 90 (2013): 53–67, https://www.sciencedirect.com/science/article/pii/S0921800913000670.

99. Jefim Vogel et al., "Socio-Economic Conditions for Satisfying Human Needs at Low Energy Use: An International Analysis of Social Provisioning," *Global Environmental Change* 69 (2021): 102287, https://www.sciencedirect.com/science/article/pii/S0959378021000662.

100. Ed Diener and Robert Biswas-Diener, "Will Money Increase Subjective Well-Being?" *Social Indicators Research* 57 (2002): 119–69, https://link.springer.com/article/10.1023/A:1014411319119.

101. Thorstein Veblen, *The Theory of the Leisure Class* (Modern Library, 1931), 103.

102. Joe Pinsker, "Why Are American Homes So Big?" *Atlantic*, September 12, 2019, https://www.theatlantic.com/family/archive/2019/09/american-houses-big/597811.

103. Jeanne Arnold et al., *Life at Home in The Twenty-First Century: 32 Families Open Their Doors* (Cotsen Institute of Archaeology, 2017).

104. Russell Belk et al., "Dirty Little Secret: Home Chaos and Professional Organizers," *Consumption Markets & Culture* 10, no. 2 (2007): 133–40.

105. Al Harris, "US Self-Storage Industry Statistics," *Sparefoot*, January 27, 2023, sparefoot.com/self-storage/news/1432-self-storage-industry-statistics.

106. James Coleman, "Social Capital in the Creation of Human Capital," *American Journal of Sociology* 94 (1988): S95–S120, https://www.journals.uchicago.edu/doi/abs/10.1086/228943; John Helliwell and Robert Putnam, "The Social Context of Well-Being," *Philosophical Transactions of the Royal Society of London B* 359, no. 1449 (2004): 1435–46.

107. Rik Pieters, "Bidirectional Dynamics of Materialism and Loneliness: Not Just a Vicious Cycle," *Journal of Consumer Research* 40, no. 4 (2013): 615–31.

108. Lilia Boujbel and Alain d'Astous, "Voluntary Simplicity and Life Satisfaction: Exploring the Mediating Role of Consumption Desires," *Journal of Consumer Behaviour* 11, no. 6 (2012): 487–94.; Tim Kasser et al., "Changes in Materialism, Changes in Psychological Well-Being: Evidence from Three Longitudinal Studies and an Intervention Experiment," *Motivation and Emotion* 38 (2014): 1–22, https://link.springer.com/article/10.1007/S11031-013-9371-4.

109. Samuel Alexander and Brendan Gleeson brought this point out in their 2018 book *Degrowth in the Suburbs*. They recommended a "material culture of sufficiency" that can not only help restrain human activity within ecologically necessary boundaries but also prepare households and localities for "disruptive and unstable economic times in which reduced consumption is enforced rather than voluntarily chosen . . . times of crisis or unplanned economic contraction." Samuel Alexander and Brendan Gleeson, "Unlearning Abundance: Suburban Practices of Energy Descent," *Degrowth in the Suburbs: A Radical Urban Imaginary* (Springer, 2018), 129.

CHAPTER III: GETTING OFF THE ROAD TO NOWHERE

110. Andre Gorz, *Ecologia* (Seagull Books, 2010), 8.1

111. Gorz, *Ecologia*, 78.

112. Chien-Yu Lin et al., "Proximity to City Centre and Cardiometabolic Risk in Middle-Aged and Older Australians: Mediating Roles of Physically Active and Sedentary Travel," *Journal of Transport & Health* 36 (2024): 101783, https://www.sciencedirect.com/science/article/pii/S221414052400029X.

113. "Cybertruck Banned from Sale in Europe: and Yet!" *Cockpit*, January 3, 2025, cockpitdz.com/en/post/cybertruck-banned-from-sale-in-europe-and-yet.

114. Gregory Shill, "Should Law Subsidize Driving?" *NYU Law Review* 95 (2020): 498–579, https://heinonline.org/hol-cgi-bin/get_pdf.cgi?handle=hein.journals/nylr95&section=13.

115. US National Transportation Safety Board, "Reducing Speeding-Related Crashes Involving Passenger Vehicles," July 25, 2017, ntsb.gov/safety/safety-studies/Documents/SS1701.pdf.

116. Shill, "Should Law."

117. Shill, "Should Law."

118. "Pedestrians Killed in Road Traffic Accidents," United Nations Economic Commission for Europe Transport Statistics Database, undated, w3.unece.org/PXWeb/en/Table?IndicatorCode=59.

119. "Dangerous by Design, 2024," Smart Growth America, smartgrowthamerica.org/dangerous-by-design.
120. John Greenfield, "Stop Victim Blaming Pedestrians and Cyclists Fatally Struck by Drivers," *Chicago Reader*, December 6, 2016, chicagoreader.com/columns-opinion/stop-victim-blaming-pedestrians-and-cyclists-fatally-struck-by-drivers; Shill, "Should Law."
121. Emma Fitzsimmons, "Jaywalking Is a New York Tradition. Now It's Legal, Too," *New York Times*, October 29, 2024, https://www.nytimes.com/2024/10/29/nyregion/jaywalking-legal-law-nyc.html; Gersh Kuntzman, "Brooklyn Pol: NYPD's Enforcement of 'Jaywalking' Is a 'Racial Injustice'," *Streetsblog NYC*, June 25, 2024, streetsblog.org/2024/06/25/brooklyn-pol-nypds-enforcement-of-jaywalking-is-a-racial-injustice.
122. Robert Schneider and Rebecca Sanders, "Pedestrian Safety Practitioners' Perspectives of Driver Yielding Behavior Across North America," *Transportation Research Record* 2519 (2015): 39–50, https://journals.sagepub.com/doi/10.3141/2519-05.
123. Courtney Coughenour et al., "Examining Racial Bias as a Potential Factor in Pedestrian Crashes," *Accident Analysis & Prevention* 98 (2017): 96–100, https://www.sciencedirect.com/science/article/pii/S000145751630361X. The authors speculated that drivers in low-income neighborhoods are accustomed to seeing more pedestrians crossing streets than those who drive mostly in affluent areas.
124. Daniel de Visé, "Pedestrian Deaths Have Risen 70 Percent Since 2010. Blame Trucks," *Hill*, April 28, 2023, https://thehill.com/policy/transportation/3976315-pedestrian-deaths-have-risen-70-percent-since-2010-blame-trucks/.
125. Grace Hauck, "Road Rage Shootings Are Increasing. There Were More Than 500 Last Year, Report Finds," *USA Today*, March 20, 2023, https://www.usatoday.com/story/news/nation/2023/03/20/road-rage-shootings-rising-us-report-finds/11484488002/.
126. "Aggressive Driving Enforcement," National Highway Traffic Safety Administration, 2004, nhtsa.gov/sites/nhtsa.gov/files/809707.pdf.
127. Jess Bidgood et al., "The Car Becomes the Weapon," *Boston Globe*, October 31, 2021, https://apps.bostonglobe.com/news/nation/2021/10/vehicle-rammings-against-protesters/tulsa/.
128. Bidgood, "The Car."
129. Sophie Clark, "Ron DeSantis Says Floridians Have Right to Hit Protesters with Cars," *Newsweek*, June 12, 2025, https://www.newsweek.com/ron-desantis-says-floridians-have-right-hit-protesters-cars-2084418.
130. Shill, "Should Law."
131. Henry Grabar, "Parking, Parking, Everywhere, but Not a Spot for Me," *Slate*, May 3, 2023, https://slate.com/business/2023/05/parking-spots-cities-paved-paradise-cars.html.
132. Dante Ramos, "How Parking Ruined Everything," *Atlantic*, July/August 2023, citing the US Department of Housing and Urban Development, https://www.theatlantic.com/magazine/archive/2023/07/cars-parking-take-up-street-space-cities/674174/.
133. Grabar, "Parking."
134. Alternatively, to spare land by erecting a multilevel garage means shelling out a ton of money. The construction of a big, new parking garage in Stamford, Connecticut, in 2023 cost an estimated $88,000 per parking space; Jared Weber,

"Stamford Transportation Center Renovations: Here's What Train Riders Need to Know," *Stamford Advocate*, My 1, 2023, https://www.stamfordadvocate.com/local/article/stamford-train-transportation-center-renovations-17918598.php

135. Grabar, "Parking."
136. Shill, "Should Law."
137. Mary Pat McGuire, "While We're Considering Removing Highways, Let's Not Overlook Pavement," *Next City*, July 7, 2021, nextcity.org/urbanist-news/while-were-considering-removing-highways-lets-not-overlook-pavement. Of those paved square miles, an area of approximately the size of Massachusetts and Connecticut, combined, is devoted just to parking vehicles: Drew Pavlick et al., "Human Health and the Transportation Infrastructure," *Journal of Human Resource and Sustainability Studies* 8, no. 3 (2020): 219–48; Grabar, *Parking*. The midpoint of Grabar's estimate range for the number of parking spaces is 1.5 billion. Multiplying that by three hundred square feet per parking space and dividing by the number of square feet per square mile results in about sixteen thousand squares miles of parking space.
138. McGuire, "While We're."
139. Jeremy Hoffman et al., "The Effects of Historical Housing Policies on Resident Exposure to Intra-Urban Heat: A Study of 108 US Urban Areas," *Climate* 8, no. 1 (2020): 12.
140. Shannon Osaka, "Will America Ever Stop Building More Highways?" *Washington Post*, February 15, 2024, https://www.washingtonpost.com/climate-solutions/2024/02/15/will-america-ever-stop-building-more-highways/.
141. Deborah N. Archer, "White Men's Roads Through Black Men's Homes: Advancing Racial Equity Through Highway Reconstruction," *Vanderbilt Law Review* 20: 1259–1330, https://heinonline.org/hol-cgi-bin/get_pdf.cgi?handle=hein.journals/vanlr73&section=36 , quoting Raymond Mohl, "The Interstates and the Cities: The U.S. Department of Transportation and the Freeway Revolt, 1966-1973," *Journal of Policy History* 20, no. 2 (2008): 193-226.
142. Archer, "White Men's Roads."
143. Avichal Mahajan, "Highways and Segregation," *Journal of Urban Economics* 141 (2024): 103574.
144. Mahajan, "Highways and Segregation."
145. "Will America," A new national group called the Freeway Fighters is bringing activists together to oppose highway expansion: America Walks, "Communities Over Highways Call for Action," February 6, 2024, americawalks.org/wp-content/uploads/2024/02/Communities-Over-Highways-Call-For-Action.pdf; Freeway Fighters Network, freeway-fighters.org.
146. Kea Wilson, "How Trump's Radical Remaking of Environmental Review Process Could Reshape Transportation," *StreetsBlog USA*, March 3, 2025, usa.streetsblog.org/2025/03/03/trumps-radical-remaking-of-environmental-review-process-will-hurt-transportation.
147. David Zipper, "The Reckless Policies that Helped Fill Our Streets with Ridiculously Large Cars," *Vox*, April 28, 2024, https://www.vox.com/future-perfect/24139147/suvs-trucks-popularity-federal-policy-pollution; Erik Shilling, "Trucks and SUVs are Now Over 80 Percent of New Car Sales in the US," *Jalopnik*, January 27, 2022, jalopnik.com/trucks-and-suvs-are-now-over-80-percent-of-new-car-sale-1848427797.

148. David Zipper, "Carry That Weight," *Slate*, June 28, 2023, https://slate.com/business/2023/06/electric-vehicles-auto-haulers-weight-capacity-roads.html.

149. Ford Motor Company advertisement, 2014, media.ford.com/content/fordmedia/fna/us/en/news/2014/01/13/ford-uses-high-strength-steel-plus-high-strength--aluminum-alloy.html; Aaron Gordon, "American Cars Are Now Almost As Big As the Tanks That Won WWII," *Vice Motherboard*, July 23, 2021, https://www.vice.com/en/article/american-cars-are-now-almost-as-big-as-the-tanks-that-won-wwii.

150. Angie Schmitt, "What Happened to Pickup Trucks?" *Bloomberg*, March 13, 2021, https://www.bloomberg.com/news/articles/2021-03-11/the-dangerous-rise-of-the-supersized-pickup-truck.

151. George Kennedy, "The Best Dually Trucks of 2024," *Car Gurus*, cargurus.com/Cars/articles/best-dually-pickup-trucks.

152. Dan Albert, "The American Pickup," *n+1*, September 7, 2016, nplusonemag.com/online-only/online-only/the-american-pickup.

153. Daniel de Visé, "Pedestrian Deaths."

154. de Visé, "Pedestrian Deaths." Death and injury rates were higher for female pedestrians: Schmitt, "What Happened."

155. Dietrich Jehle et al., "Car Ratings Take a Back Seat to Vehicle Type: Outcomes of SUV Versus Passenger Car Crashes," *HCA Healthcare Journal of Medicine* 2, no. 4 (2021): 289–95. The high front end of today's pickups and SUVs also blocks the driver's view of anything that's situated ahead of the truck, low, and close. This has led to increased deaths among children. In tragic "frontover" incidents, drivers pulling the vehicle forward, often in a driveway, fail to see that there's a child in their path. When some observers expressed incredulity that a parent could fail to notice their own child right in front of them, a group of Virginia parents gave the NBC4 TV station in Washington, DC, permission to run a demonstration. One by one, children sat down in a line, starting at a full-size SUV's front bumper and continuing forward. The queue grew to ten children before the adult in the driver's seat could see the frontmost child. Susan Hogan et al., "Driveway Danger: Kids Being Injured and Killed in 'Frontover' SUV Blind Zone Incidents," *NBC Washington*, July 28, 2022, nbcwashington.com/investigations/driveway-danger-kids-being-injured-and-killed-in-frontover-suv-blind-zone-incidents/3119237.

156. David Zipper, "Car Companies Are Making a Deadly Mistake with Electric Vehicles," *Slate*, August 31, 2022, https://slate.com/technology/2022/08/electric-trucks-cars-too-heavy-inflation-reduction-act.html.

157. Zipper, "Car Companies."

158. David Zipper, "Electric Vehicles Are Bringing Out the Worst in Us," *Atlantic*, January 4, 2023, https://www.theatlantic.com/ideas/archive/2023/01/electric-vehicles-suv-battery-climate-safety/672576/.

159. EPA categorizes hydroelectric power as renewable, but for many reasons it is not ecologically friendly and has a big greenhouse gas footprint, so I am lumping it in with nonrenewable sources.

160. Pabitra Kumar Das et al., "Life Cycle Assessment of Electric Vehicles: a Systematic Review of Literature," *Environmental Science and Pollution Research* 31, no. 1 (2024): 73–89.

161. Stan Cox and Priti Gulati Cox, "Electric Vehicles Won't Save Us," *Nation*, October 18, 2022, https://www.thenation.com/article/environment/electric-vehicles-lithium-cobalt-sustainable/.

162. David Zipper, "EVs Are Sending Toxic Tire Particles into the Water, Soil, and Air," *Atlantic*, July 19, 2023, https://www.davidzipper.com/writing-2/ev-tires; Caroline Brogan, "Prioritize Tackling Toxic Emissions from Tires, Urge Imperial Experts," *Imperial News*, February 23, 2023, imperial.ac.uk/news/243333/prioritise-tackling-toxic-emissions-from-tyres/.

163. For example, there have been efforts aimed at addressing excessive parking space (a serious problem, I think we can all agree—except when *you and I* need to park). Economists can demonstrate, theoretically, that reducing the amount of free parking and adjusting parking fees to balance supply with demand would solve the problem. And indeed, commercial areas that follow such advice may well optimize their parking availability. Such optimized areas, however, can attract new influxes of motorists who are frustrated by parking hassles in other parts of town, so congestion resurges, and quality of life suffers. Raising fees yet again to more strongly discourage parking will favor the most well-off drivers, magnifying the injustice. Grabar, "Parking."

164. Here's another half measure: Done today, repeal of the federal regulatory and tax-related incentives that have fueled the light truck boom would probably be too little, too late to have much effect on vehicle manufacturers and buyers. They've been immersed in bigger-is-better automotive culture for more than two decades, and besides, if buyers trade down to a sedan, they'll be sitting ducks in the battle of the streets. In yet another example, less permissive speed-limit laws and more conscientious enforcement would doubtless save lives of both pedestrians and car occupants, but it will put hardly a dent in automotive supremacy writ large.

165. Keri Schwartz, "The Meaning Behind the Song: Air Travel—Live by Jerry Seinfeld," *BeatCrave*, beatcrave.com/w2/the-meaning-behind-the-song-air-travel-live-1998-broadhurst-theatre-by-jerry-seinfeld.

166. Leon James, "Air Rage: Bad Behavior at 30,000 Feet," *Conversation*, December 22, 2014, https://theconversation.com/air-rage-bad-behavior-at-30-000-feet-35240.

167. Brad Broberg, "UW Research Explores Links Between Air Travel and Stress," *University of Washington Magazine*, March 1, 2000, washington.edu/news/2002/03/21/number-of-passengers-experiencing-air-travel-stress-jumps-to-81-percent.

168. Broberg, "UW Research."

169. Alex Abad-Santos, "'Calculated Misery'": How Airlines Profit from Your Miserable Flying Experience," *Vox*, April 17, 2017, https://www.vox.com/culture/2017/4/14/15275642/united-airlines-calculated-misery-dragging-man-off-plane.

170. Whizy Kim, "What a Summer of Hellish Flights Taught Us About Flying Now," *Vox*, September 8, 2023, https://www.vox.com/money/23862850/flights-travel-delays-fees-nightmares-compensation-airlines.

171. "New Survey: Nearly 90 Percent of Americans Have Flown Commercially," Airlines for America, March 22, 2023, airlines.org/new-survey-nearly-90-percent-of-americans-have-flown-commercially; Lucia Binding, "'Wealthy' Minority Responsible for Majority of Global Air Travel, Says Climate Group Study," *Sky News*, March 31, 2021, news.sky.com/story/wealthy-minority-responsible-for-majority-of-global-air-travel-says-climate-group-study-12261620; Lisa Hopkinson and Sally Cairns, "Elite Status: Global Inequalities in Flying," *Possible*, March 2021, creativecommons.org/licenses/by-nc-nd/2.0/uk/.

172. Sarven McLinton et al. "'Air Rage': A Systematic Review of Research on Disruptive Airline Passenger Behaviour 1985–2020," *Journal of Airline and Airport Management* 10, no. 1 (2020): 31–49.

173. Irie Sentner, "Air Travel Has Gotten More Violent. Flight Attendants Want Training to Fight Back," *Politico*, August 20, 2023, https://www.politico.com/news/2023/08/20/flight-attendants-training-union-00111864; Francesca Street, "Unruly Passengers Were a Problem Before the Pandemic. Now They're Even Worse," CNN, January 24, 2024; https://www.cnn.com/travel/unruly-airplane-passengers-post-pandemic-problem.

174. Daniel Coyle et al. "Descriptive Analysis of Air Rage Incidents Aboard International Commercial Flights, 2000–2020," *Transportation Research Interdisciplinary Perspectives* 11 (2021): 100418, https://www.sciencedirect.com/science/article/pii/S259019822100124X.

175. Maia Szalavitz, "Air Rage: Why Does Flying Make Us So Angry? Science Says It's About Class," *Guardian*, May 25, 2017, https://economichardship.org/2017/05/air-rage-why-does-flying-make-us-so-angry-science-says-its-about-class/.

176. David Owen, *The Conundrum* (Riverhead, 2011), 94–96.

177. Owen, *Conundrum*, 95.

178. Owen, *Conundrum*, 94–95.

179. Bill Moyer et al., *Solutionary Rail: a People-Powered Campaign to Electrify America's Railroads and Open Corridors to a Clean Energy Future* (Backbone Campaign, 2016); also see solutionaryrail.org.

180. Justin Franz, "Boom in Freight Breaking Amtrak's Empire Builder," *Missoulian*, August 16, 2014, missoulian.com/news/state-and-regional/boom-in-freight-breaking-amtrak-s-empire-builder/article_7f78dd58-25c6-11e4-96fe-0014bcf887a.html; Mike Lee, "Amtrak's Dream of Speedy Service Stuck Behind Lumbering Freight Trains," *E&E News*, December 12, 2023, eenews.net/articles/amtraks-dream-of-speedy-service-stuck-behind-lumbering-freight-trains.

181. Amtrak, "Delayed by Freight: Measuring On-Time Performance Across Our Network," April 2024, amtrak.com/on-time-performance.

182. Luz Laso, "With Grant, Amtrak Plans Gulf Coast Route's Return, 18 Years After Katrina," *Washington Post*, September 25, 2023, https://www.washingtonpost.com/transportation/2023/09/25/amtrak-gulf-coast-rail-katrina; "tooth and rail" typo intended.

183. No need for high-speed rail. Existing rail lines across North America can be refurbished and extended to carry modest-speed passenger traffic following a plan like the one urged today by the Solutionary Rail campaign; Moyer et al., *Solutionary Rail*.

CHAPTER IV: THIS LAND

184. James Davis et al., "Rural America at a Glance: 2023 Edition," USDA Economic Research Service, November, 2023; "US Farms: Numbers, Size, and Ownership," USDA Economic Research Service, November 15, 2023, ers.usda.gov/publications/pub-details/?pubid=107837; "Structure and Finances of US Farms: 2005 Family Farm Report / EIB-12," USDA Economic Research Service, ers.usda.gov/webdocs/publications/43810/29468_eib12c_002.pdf?v=1268.8; Campbell Gibson, "American Demographic History Chartbook: 1790 to 2010," 2015, demographicchartbook.com/index.php/chapter-2-urban-rural-and-farm-population-and-large-cities/2-1-population-by-urban-rural-and-farm-residence-for-the-united-states-1790-

to-2010/; three percent of Americans lived on farms at the time of the 2000 US Census—the last year when it collected such data. The number of farms declined in the subsequent quarter century, so the share of the population living on farms is probably less than 3 percent now. For number of farms in 2002, see "Structure."

185. Timothée Parrique, "The Political Economy of Degrowth" (PhD diss., Stockholm University, 2019), 516. As one small piece of his epic work, Parrique collected 540 elements of the degrowth vision submitted to the French government from citizens across the nation during its 2019 Great National Debate. Of these 540 goals, objectives, and proposed policies, Parrique found 86 pertaining to agriculture.

186. Stan Cox, "The Wayward Carbon Atoms of the Anthropocene," *Carbon Copy* 3 (2022), carboncopy.world/cox.

187. "Iowa Ag News," National Agricultural Statistics Service, January 12, 2024, nass. usda.gov/Statistics_by_State/Iowa/Publications/Crop_Report/2024/IA-Crop-Pro-duction-Annual-01-24.pdf.

188. "Acreage," US Department of Agriculture, National Agricultural Statistics Service, June 30, 2021, nass.usda.gov/Publications/Todays_Reports/reports/acrg0621.pdf.

189. S.R. Srocco, "The US Corn Ethanol Boondoggle," *Investment Watch*, July 27, 2019, investmentwatchblog.com/the-u-s-corn-ethanol-boondoggle-producing-1-million-barrels-per-day-of-unprofitable-energy/; Tyler Lark et al., "Environmental Outcomes of the US Renewable Fuel Standard," *Proceedings of the National Academy of Sciences* 119 (2022): e2101084119.

190. John Schramski et al., "Energy Use and the Sustainability of Intensifying Food Production," *Nature Sustainability* 3, no. 4 (2020): 257–59.

191. Carrie Hribar, "Understanding Concentrated Animal Feeding Operations and Their Impact on Communities," National Association of Local Boards of Health, 2010, stacks.cdc.gov/view/cdc/59792; Sarah Porter and Craig Cox, "Manure Overload: Manure Plus Fertilizer Overwhelms Minnesota's Land and Water," Environmental Working Group, May 28, 2020, ewg.org/interactive-maps/2020-ma-nure-overload.

192. Will Potter, *Little Red Barns: Hiding the Truth from Farm to Fable* (City Lights, 2025), 112-117.

193. Potter, *Little Red Barns*, 119.

194. Hribar, "Understanding."

195. Melina Walling and Michael Phillis, "Latest EPA Assessment Shows Almost No Improvement in River and Stream Nitrogen Pollution," Associated Press, January 21, 2024; Soren Rundquist, "EWG Water Atlas Reveals Nitrate, Phosphorous Water Pollution in Four Upper Mississippi Basin States Closely Aligns with Fertilizer Use on Cropland," Environmental Working Group, August 31, 2021, ewg.org/research/ewg-water-atlas-reveals-nitrate-phosphorous-water-pollution-four-up-per-mississippi-basin; Natalie Krebs, "Nitrate levels Are Often Higher in the Rural Midwest. How Does This Affect Health?" Iowa Public Radio, November 1, 2023, iowapublicradio.org/health/2023-11-01/nitrate-levels-are-often-higher-in-the-rural-midwest-how-does-this-affect-health; Yanqui Xu, "Our Dirty Water: Nebraska's Nitrate Problem Is Growing Worse," *Investigate Midwest*, October 31, 2022, investigatemidwest.org/2022/10/31/our-dirty-water-nebraskas-nitrate-prob-lem-is-growing-worse-its-likely-harming-our-kids; Greg Stanley, "A Minnesota Agency Was Supposed to Limit Nitrates a Decade Ago. Officials Say They Can't,"

Star Tribune, March 4, 2023, startribune.com/nitrate-pollution-limits-mpca-directive-from-minnesota-legislature-environment-water-fish/600256304.

196. Madison McVan, "OSHA Should Strengthen Rules to Protect Meat and Poultry Workers from Infectious Disease, Government Watchdog Says," *Investigate Midwest*, June 22, 2023, investigatemidwest.org/2023/06/22/osha-should-strengthen-rules-to-protect-meat-and-poultry-workers-from-infectious-disease-government-watchdog-says.

197. "Advanced Food Manufacturing," National Institute of Food and Agriculture, nifa.usda.gov/topics/advanced-food-manufacturing.

198. Monica Crippa et al., "Food Systems Are Responsible for a Third of Global Anthropogenic GHG Emissions," *Nature Food* 2, no. 3 (2021): 198–209.

199. June Sekera and Andreas Lichtenberger, "Assessing Carbon Capture: Public Policy, Science, and Societal Need," *Biophysical Economics and Sustainability* 5, no, 14 (2020): 1–28.

200. Stan Cox, "'We'll Meet Them Out in the Fields'—Challenging the Pipelines to Nowhere," *Resilience*, September 16. 2022, resilience.org/stories/2022-09-16/well-meet-them-out-in-the-fields-challenging-the-pipelines-to-nowhere.

201. Jared Strong, "Summit says Pipeline System Won't Be Operational Until 2026," *Nebraska Examiner*, October 19, 2023, nebraskaexaminer.com/briefs/summit-says-pipeline-system-wont-be-operational-until-2026/.

202. A. Bekkerman et al., "Does Farm Size Matter? Distribution of Crop Insurance Subsidies and Government Program Payments Across US Farms," *Applied Economic Perspectives and Policy* 41, no. 3 (2019): 498–518.

203. Alyssa Casey, "Racial Equity in US Farming," Congressional Research Service, November 19, 2021, crsreports.congress.gov/product/pdf/R/R46969. Also in the article: "The proportion of US farmers who identify as American Indian or Alaska Native increased from less than 1% in 1900 to 2.3% in 2017. Asian farmers remained roughly constant during the same period at less than 1% of farmers. Farmers of Spanish, Hispanic, or Latino ethnicity increased from less than 1% of all farmers in 1920 to 3.4% in 2017."

204. Amy Mayer, "Can $3 Billion Persuade Black Farmers to Trust the Department of Agriculture?" NPR, January 8, 2024, npr.org/2023/12/26/1221725620/where-biden-administration-3-billion-farming-grant-has-been-going; "USDA Provides $208 Million to Help Prevent Guaranteed Borrower Foreclosures and to Assist Emergency Loan Borrowers," USDA press release, November 30, 2023, usda.gov/media/press-releases/2023/11/30/usda-provides-208-million-help-prevent-guaranteed-borrower. In 2022, the Agriculture Department announced the $1 billion Partnerships for Climate-Smart Commodities granting program that was to include provisions for Black, Indigenous, and other farmers regarded as historically underserved. Mayer reported, however, that, "many projects receiving the most money are run by giant for-profit companies and major agricultural lobbying groups that don't appear to have a clear plan for how they will serve disadvantaged farmers." Furthermore, "the USDA's definition of 'historically underserved' farmers includes not only ethnic and racial minorities and women but also veterans, young and beginning farmers, and those operating at poverty level—so it's possible for a project to meet the USDA's equity goal without serving any Black farmers at all." (Mayer, "Can $3 billion?")

205. Mayer, "Can $3 Billion?"

206. Gosia Wozniacka, "Indigenous Leaders Call for Landback Reforms and Climate Justice in 'Required Reading,'" *Yes! Magazine*, November 5, 2021, https://www.yesmagazine.org/environment/2021/11/05/indigenous-authors-land-back-climate-justice. From the article, here's Kailea Frederick, a climate justice organizer for NDN Collective, on the link between two of NDN's campaigns, Landback and climate: "We need land returned in large quantities at this moment so that Indigenous peoples can be in direct conversation with the land and engaged in their traditional practices, which inherently mitigate climate change."

207. Woznizcka, "Indigenous Leaders."

208. "Coming Together for Land Co-Management," Native American Rights Fund, February 5, 2025, narf.org/shared-horizons-comanagement-conference.

209. "CSKT Bison Range," bisonrange.org.

210. Jim Robbins, "How Returning Lands to Native Tribes Is Helping Protect Nature," *Yale E360*, June 3, 2021, e360.yale.edu/features/how-returning-lands-to-native-tribes-is-helping-protect-nature.

211. Sophie Austin, "California Tribes Will Manage, Protect State Coastal Areas," Associated Press, October 9, 2022, https://apnews.com/article/california-sacramento-gavin-newsom-climate-and-environment-government-politics-f3a37f53129ef43ffd60f9b58506247b.

212. Andrew Staples, "Ancestral Land in Butte Creek Canyon Returned to the Mechoopda Tribe," *Chico State Today*, September 23, 2022, today.csuchico.edu/bcep-transferred-to-mechoopda-tribe; "Central Valley Spring-Run Chinook Salmon, *NOAA Fisheries*, July 1, 2024, fisheries.noaa.gov/west-coast/endangered-species-conservation/central-valley-spring-run-chinook-salmon.

213. Vanessa Racehorse and Anna Hohag, "Achieving Climate Justice Through Land Back: An Overview of Tribal Dispossession, Land Return Efforts, and Practical Mechanisms for #Landback," *Colorado Natural Resources Energy & Environment Law Review* 34 (2023): 175–212.

214. Racehorse, "Achieving Climate Justice."

215. Robbins, "How Returning"; Alice Hutton, "Native American Tribe in Maine Buys Back Island Taken 160 Years Ago," *Guardian*, June 4, 2021, theguardian.com/us-news/2021/jun/04/native-american-tribe-maine-buys-back-pine-island.

216. *Third Annual Report on Tribal Co-Stewardship* (Department of the Interior, December 5, 2024), doi.gov/media/document/annual-tribal-co-stewardship-report-2024-12-05; "SO 3416 - Ending DEI Programs and Gender Ideology Extremism," Secretarial Order, Department of the Interior, January 30, 2025, doi.gov/document-library/secretary-order/so-3416-ending-dei-programs-and-gender-ideology-extremism ; "Interior Department Issues Guidance to Strengthen Tribal Co-Stewardship of Public Lands and Waters," press release, Department of the Interior, September 13, 2022 (as revised in January, 2024); since January 20, 2025 the Interior Department web page for the press release carries a notice that the co-stewardship program has been effectively mothballed: doi.gov/pressreleases/interior-department-issues-guidance-strengthen-tribal-co-stewardship-public-lands-and.

217. Parrique, *Political Economy*, 851–2.

218. Stan Cox, *The Path to a Livable Future* (San Francisco: City Lights, 2010), 30–34.

219. Daniel Costa, "The Farmworker Wage Gap," Economic Policy Institute, October 5, 2023, epi.org/blog/the-farmworker-wage-gap-farmworkers-earned-40-less-than-comparable-nonagricultural-workers-in-2022. Out in the fields, writes Costa, farmwork-

ers can face even deadlier temperatures and are 35 times as likely to die from heat stress as laborers in other industries; Cox, *The Path*, 109–111.

220. Tristan Roberts, "We Spend 90% of Our Time Indoors. Says Who?" *Building-Green*, December 15, 2016, buildinggreen.com/blog/we-spend-90-our-time-in-doors-says-who.

221. Richard Louv, *The Last Child in the Woods: Saving Our Children from Nature Deficit Disorder* (Algonquin, 2008).

222. Louv, *The Last Child*.

223. Petra Pfefferle, Corinna Keber, Robert Cohen, and Holger Garn, "The Hygiene Hypothesis–Learning from but Not Living in the Past," *Frontiers In Immunology* 12 (2021): 635935, https://www.frontiersin.org/articles/10.3389/fimmu.2021.635935/full.

224. Harry Stevens, "Mapping America's Access to Nature, Neighborhood by Neighborhood," *Washington Post*, April 10, 2024, https://www.washingtonpost.com/climate-environment/interactive/2024/nature-health-maps-neighborhood-city/. Such racial disparities in access to non-human nature are no accident. For example, from the 1930s to the 1960s, predominantly Black US neighborhoods were discriminated against in home-mortgage lending, in a practice known as "redlining." These areas continue to be economically depressed more than a half-century later, and people of color still in the majority. Formerly redlined neighborhoods are also home to far more concrete and asphalt and less green space than non-redlined areas in the same city: Hoffman, "Effects of Historical Housing.

225. Jacqueline L. Scott and Ambika Tenneti, "Race and Nature in the City: Engaging Youth of Color in Nature-Based Activities," Nature Canada, naturecanada.ca/race-and-nature-in-the-city.

226. Cassandra Johnson Gaither et al., ""Black Folks Do Forage": Examining Wild Food Gathering in Southeast Atlanta Communities," *Urban Forestry & Urban Greening* 56 (2020): 126860, https://www.sciencedirect.com/science/article/pii/S1618866720306774.

227. Meg St-Esprit McKivigan, "'Nature Deficit Disorder' Is Really a Thing," *New York Times*, June 23, 2020, https://www.nytimes.com/2020/06/23/parenting/nature-health-benefits-coronavirus-outdoors.html.

228. Wes Jackson and Wendell Berry, "A 50-Year Farm Bill," *New York Times*, Jan 5, 2009, https://landinstitute.org/media-coverage/ny-times-op-ed-piece-wendell-berry-wes-jackson/; The Land Institute, "A 50-Year Farm Bill," landinstitute.org/wp-content/uploads/2016/09/FB-edited-7-6-10.pdf.

229. Wes Jackson, "Where We Are Going," *The Land Report* 76 (2003): 16, landinstitute.org/wp-content/uploads/2018/05/96642-LR-76.pdf.

230. Thays Millena Alves Pedroso et al., "Cancer and Occupational Exposure to Pesticides: A Bibliometric Study of the Past 10 years," *Environmental Science and Pollution Research* (2022): 1–12, https://link.springer.com/article/10.1007/s11356-021-17031-2.

231. Wes Jackson says, "If you have just the family farm, and your community is gone, then money comes through the hands of the farmer and goes to the suppliers of inputs, the people who supply the pesticides, the fertilizers, the farm machinery and so on—effectively the bankers of the farmer—and then it gets bounced electronically around the world, wherever the interest rate happens to be high. On the other hand, if it came through the farmer and there was a minimum of

these inputs, in other words more of that which comes from nature rather than that which is provided by industry, then the money goes through the hands of the farmer into the local community, the small town or whatever. Then you can sponsor the rural churches, schools, baseball, all of the stuff that goes to make up rural culture." Interview with Robert Gilman, "Mainstreaming Sustainable Agriculture," *Sustainable Habitat*, Autumn, 1986, context.org/iclib/ic14/wjackson.

232. Anousheh Nikkhou and Azime Tezer, "Nature-Deficit Disorder in Modern Cities," in *Sustainable Development and Planning XI*, eds. S. Syngellakis and S. Hernández (WIT Press, 2020): 407–17; Evelyn N. Alvarez, Alexys Garcia, and Pauline Le, "A Review of Nature Deficit Disorder (NDD) and Its Disproportionate Impacts on Latinx Populations," *Environmental Development* 43 (2022): 100732, https://www.sciencedirect.com/science/article/pii/S2211464522000343

233. Allison Karpyn et al., "The Changing Landscape of Food Deserts," *UNSCN Nutrition* 44 (2019): 46, https://pmc.ncbi.nlm.nih.gov/articles/PMC7299236/; Armita Kar et al., "COVID-19 Exacerbates Unequal Food Access," *Applied Geography* 134 (2021): 102517, https://www.sciencedirect.com/science/article/pii/S0143622821001338; Bethan Moorcraft, "'Power to Communities': Chicago Considers City-Owned Grocery Store to Address 'Food Deserts' After Giants Like Walmart and Whole Foods Shutter Stores," *Moneywise*, September 19, 2023, moneywise.com/news/economy/chicago-explores-city-owned-grocery-store.

CHAPTER V: TAKE MY JOB—PLEASE!

234. David Graeber, "The Modern Phenomenon of Bullshit Jobs," reprinted by the *Sydney Morning Herald*, September 3, 2013, smh.com.au/public-service/the-modern-phenomenon-of-bullshit-jobs-20130831-2sy3j.html.

235. David Graeber, *Bullshit Jobs: A Theory* (Simon and Schuster, 2018), 8.

236. Graeber, *Bullshit Jobs*, 27.

237. Michele Hellebuyck et al., "Mind the Workplace," Mental Health America, 2017, archive.hshsl.umaryland.edu/handle/10713/14084; Douglas LaBier, "Most People Are Unhappy with Their Jobs, New Survey Finds," *Progressive Impact*, December 5, 2017, progressiveimpact.org/most-people-are-unhappy-with-their-jobs-new-survey-finds/.

238. Caitlin Mullen, "There's a Growing Gap in Job Satisfaction," *Business Journals*, October 30, 2019, bizjournals.com/bizwomen/news/latest-news/2019/10/theres-a-growing-gap-in-job-satisfaction.html.

239. I checked the articles out, and they all seemed to offer straightforward, practical advice; none were obviously AI-generated. But they didn't address a bigger question: How and why the creation of so many jobs—activities in which most adults spend a large share of their waking hours—that inspire such loathing?

240. Katherine Haan, "Gender Pay Gap Statistics in 2024," *Forbes*, March 1, 2024, forbes.com/advisor/business/gender-pay-gap-statistics/#general_gender_wage_gap_statistics_section.

241. Cynthia Hess et al., "Providing Unpaid Household and Care Work in the United States: Uncovering Inequality," Institute for Women's Policy Research, Briefing Paper #C487, January 2020, iwpr.org/wp-content/uploads/2020/01/IWPR-Providing-Unpaid-Household-and-Care-Work-in-the-United-States-Uncovering-Inequality.pdf.

242. Timothée Parrique, *The Political Economy of Degrowth*, 584.

243. Juliana Kaplan, "Almost 70% of Workers Want a Career Change. They'd Take Better Work-Life Balance over Higher Pay," *Business Insider*, August 16, 2021, https://www.businessinsider.com/workers-want-work-life-balance-more-than-higher-pay-2021-8?op=1.

244. David Spencer, "Four-Day Work Week Is a Necessary Part of Human Progress—Here's a Plan to Make It Happen," *Conversation*, September 24, 2019, https://theconversation.com/four-day-work-week-is-a-necessary-part-of-human-progress-heres-a-plan-to-make-it-happen-124104.

245. Anna Coote et al., *21 Hours: Why a Shorter Working Week Can Help Us All to Flourish in the 21st Century* (New Economics Foundation, 2010), https://neweconomics.org/2010/02/21-hours/.

246. Douglas Broom, "Four-Day Work Week Trial in Spain Leads to Healthier Workers, Less Pollution," World Economic Forum, October 25, 2023, weforum.org/agenda/2023/10/surprising-benefits-four-day-week.

247. Orla Kelly et al., "The Four Day Week: Assessing Global Trials of Reduced Work Time with No Reduction in Pay: Evidence from Ireland," University College Dublin, 2022, researchrepository.ucd.ie/handle/10197/25259.

248. Jan Dönges and Sophie Bushwick, "A Four-Day Workweek Reduces Stress Without Hurting Productivity," *Scientific American*, March 7, 2023, https://www.scientificamerican.com/article/a-four-day-workweek-reduces-stress-without-hurting-productivity.

249. Zinnia Maldonado, "Companies Around the World Adopt Four-Day Work Week Pilot Programs to Meet Growing Demand from Gen-Z and Millennial Employees," *CBS News*, January 2, 2024, https://www.cbsnews.com/newyork/news/companies-around-the-world-adopt-four-day-work-week-pilot-programs-to-meet-growing-demand-from-gen-z-and-millennial-employees/ But only 13 percent liked the idea of a four-day week *with a pay cut*.

250. Broom, "Four-Day."

251. Campbell, Timothy T., "The Four-Day Work Week: A Chronological, Systematic Review of the Academic Literature'" *Management Review Quarterly* 74, no. 3 (2023): 1791–1807.

252. Ibid, emphasis mine.

253. Sarah Jaffe, "A 32-Hour Workweek Is Ours for the Taking," *In These Times*, April 2, 2024, https://inthesetimes.com/article/unions-4-day-work-week-uaw-covid-strike-demands-labor; Chris Isidore, "The 4-Day Workweek Was a Longshot. The UAW Isn't Giving Up," CNN, December 16, 2023, https://www.cnn.com/2023/12/16/business/shawn-fain-4-day-work-week.

254. Jane Their, "If You Want a 3-Day Weekend, Work at One of These Companies," *Fortune*, October 5, 2022, https://finance.yahoo.com/news/want-3-day-weekend-one-194843851.html.

255. Coote et al., *21 Hours*. They write, "Shorter working hours could help to adapt the economy to the needs of society and the environment, rather than subjugating society and environment to the needs of the economy. Business would benefit from more women entering the workforce; from men leading more rounded, balanced lives; and from reductions in work-place stress associated with juggling paid employment and home-based responsibilities. It could also help to end credit-fueled growth, to develop a more resilient and adaptable economy, and to

safeguard public resources for investment in a low-carbon industrial strategy and other measures to support a sustainable economy."

256. The twenty-hour workweek features prominently in Ernest Callenbach's somewhat degrowthy 1974 novel *Ecotopia: Thirtieth Anniversary Edition* (Banyan Tree, 2004).

257. Callenbach, *Ecotopia*.

258. Kyle Knight et al., "Could Working Less Reduce Pressures on the Environment? A Cross-National Panel Analysis of OECD Countries, 1970–2007," *Global Environmental Change* 23, no. 4 (2013): 691–700.

259. St. Louis Federal Reserve, "Value Added by Industry: Finance, Insurance, Real Estate, Rental, and Leasing as a Percentage of GDP," June 29, 2023, fred. stlouisfed.org/series/VAPGDPFIRL and "Value Added by Industry as a Percentage of Gross Domestic Product," June 27, 2024, fred.stlouisfed.org/release/tables?rid=331&eid=211#snid=875.

260. Graeber, *Bullshit Jobs*, 165.

261. Ibid, 9–58.

262. Zielinska-Dabkowska et al., "Reducing Nighttime Light Exposure in the Urban Environment to Benefit Human Health and Society," *Science* 380, no.6650 (2023): 1130–35; "Job Flexibilities and Work Schedules Summary," Bureau of Labor Statistics, 2019, bls.gov/news.release/flex2.nro.htm.

263. Jesús Rodrigo-Comino et al., "Light Pollution: A Review of the Scientific Literature," *The Anthropocene Review* 10, no. 2 (2023): 367–92; Zielinska-Dabkowska et al., "Reducing"; Amedeo Argentiero et al., "Outdoor Light Pollution and COVID-19: The Italian Case," *Environmental Impact Assessment Review* 90 (2021): 106602, https://www.sciencedirect.com/science/article/pii/S0195925521000524; Ela Rydz et al., "Prevalence and Recent Trends in Exposure to Night Shiftwork in Canada," *Annals of Work Exposures and Health* 64, no. 3 (2020): 270–81; Hannah Dreier and Meridith Kohut, "The Kids on the Night Shift," *New York Times*, September 18, 2023, https://www.nytimes.com/2023/09/18/magazine/child-labor-dangerous-jobs.html; Juliana Kim, "Perdue Farms and Tyson Foods Under Federal Inquiry over Reports of Illegal Child Labor," NPR, September 25, 2023, https://www.npr.org/2023/09/25/1201524399/child-labor-perdue-farms-tyson-foods-investigation; Katie Wermus, "Minnesota Children Illegally Hired to Clean Slaughterhouses: Federal Lawsuit," *FOX 9/KMSP*, November 11, 2022, fox9.com/news/minnesota-minors-were-illegally-hired-to-clean-slaughterhouses-labor-dept-says; Remy Tumin, "Labor Department Finds 31 Children Cleaning Meatpacking Plants," *New York Times*, November 11, 2022, https://www.nytimes.com/2022/11/11/business/child-labor-meatpacking-plants.html.

264. Shawna Nadybal et al., "Light Pollution Inequities in the Continental United States: A Distributive Environmental Justice Analysis," *Environmental Research* 189 (2020): 109959, https://www.sciencedirect.com/science/article/pii/S0013935120308549; health risks of night work for workers younger than twenty-five years, however, were lower than for older workers: Cody Ramin et al., "Night Shift Work at Specific Age Ranges and Chronic Disease Risk Factors," *Occupational and Environmental Medicine* 72, no.2 (2015): 100–107.

265. Graeber classifies as "duct-taping" the subset of such positions that exist only for the employee to fix problems created by other employees of the same company *as part of their job*; Graeber, *Bullshit Jobs*, 41–44.

266. Susie Khamis, "The Aestheticization of Restraint: The Popular Appeal of De-Cluttering After the Global Financial Crisis," *Journal of Consumer Culture* 19, no. 4 (2019): 513–31.
267. Russell Belk et al., "Dirty Little Secret: Home Chaos and Professional Organizers," *Consumption, Markets and Culture* 10, no. 2 (2007): 133–40.
268. Belk et al., "Dirty Little Secret."
269. Khamis, "Aestheticization."
270. Javier Lloveras et al., "On 'the Politics of Repair Beyond Repair': Radical Democracy and the Right-to-Repair Movement," *Journal of Business Ethics* 196, no. 2 (2025): 325-344.
271. Lloveras et al., "On 'the Politics"; National Conference of State Legislatures, "Right to Repair 2023 Legislation," November 1, 2023, ncsl.org/technology-and-communication/right-to-repair-2023-legislation.
272. "Military Careers," US Bureau of Labor Statistics, www.bls.gov/ooh/military/military-careers.htm; Alan Ott, "Defense Primer: Department of Defense Civilian Employees," Congressional Research Service, February 6, 2023, sgp.fas.org/crs/natsec/IF11510.pdf.
273. David Vine, *Base Nation: How US Military Bases Abroad Harm America and the World* (Metropolitan, 2015).
274. David Vine, "The United States Probably Has More Foreign Military Bases Than Any Other People, Nation, or Empire in History," *Nation*, September 14, 2015, https://www.thenation.com/article/world/the-united-states-probably-has-more-foreign-military-bases-than-any-other-people-nation-or-empire-in-history/.
275. Patrick Bigger et al., "The Carbon Bootprint of the US Military and Prospects for a Safer Climate," in *Negotiating Climate Change in Crisis*, eds. Steffen Böhm and Sian Sullivan (Open Book Publishers, 2021), 53–62.
276. Neta C. Crawford, "*Pentagon Fuel Use, Climate Change, and the Costs of War*," Boston University, November 13, 2019, watson.brown.edu/costsofwar/files/cow/imce/papers/Pentagon%20Fuel%20Use%2C%20Climate%20Change%20and%20the%20Costs%20of%20War%20Revised%20November%202019%20Crawford.pdf.
277. Sara Mar, "Indigenous People of Guam Are Fighting US Militarism and Environmental Ruin," *Truthout*, May 31, 2024, https://truthout.org/articles/indigenous-people-of-guam-are-fighting-us-militarism-and-environmental-ruin/.
278. Ann Tyson, "Youths in Rural US Are Drawn to Military," *Washington Post*, November 4, 2005, https://www.washingtonpost.com/archive/politics/2005/11/04/youths-in-rural-us-are-drawn-to-military/24122550-6bb7-4174-93a0-7e1d91a78b2d : "Many of today's recruits are financially strapped, with nearly half coming from lower-middle-class to poor households."
279. Beth Asch et al., *Food Insecurity Among Members of the Armed Forces and Their Dependents* (RAND Corporation, January 3, 2023), rand.org/pubs/research_reports/RRA1230-1.html.
280. James Hanlon and Orion Donovan Smith, "A Quarter of Military Families Struggle to Afford Enough Food. Northwest Lawmakers Are Trying to Help," *Spokesman-Review*, February 18, 2024, https://www.spokesman.com/stories/2024/feb/18/a-quarter-of-military-families-struggle-to-afford-/; Asch et al., "Food Insecurity."
281. Asch et al., "Food Insecurity."

282. Sarah Sicard, "How Much Land Does the Military Really Own?" *Military Times*, Aug 15, 2022, https://www.militarytimes.com/off-duty/military-culture/2022/08/15/how-much-land-does-the-military-really-own.

283. John Hamilton, "Contamination at US Military Bases: Profiles and Responses," *Stanford Environmental Law Journal* 35 (2016): 223–49, https://heinonline.org/hol-cgi-in/get_pdf.cgi?handle=hein.journals/staev35&section=10.

284. Christopher Rein, "The Environmental Impact of the US Army Air Forces' Production and Training Infrastructure on the Great Plains" in *Airpower and the Environment: The Ecological Implications of Modern Air Warfare*, ed. Joel Hayward (Air University Press, 2013), 25–42.

285. Jason Beets, "Salina Environmental Cleanup to Cost $95 Million," *Salina Journal*, May 2, 2019, https://www.salina.com/story/news/local/2019/05/02/salina-environmental-clean-up-to-cost-95-million/5259581007/; Jeremy Bohn, "$65.9M Settlement Reached in Environmental Cleanup at Former Schilling Air Force Base," KSAL, June 24, 2020, ksal.com/65-9m-settlement-reached-in-environmental-cleanup-at-schilling-air-force-base; Tim Unruh, "Schilling Cleanup: It Is Time to 'Turn Some Dirt'," Salina Airport Authority press release, April 27, 2021, salinaairport.com/media/36606/schilling-cleanup.html.

286. Sharon Lerner and Lisa Song, "TCE Is Linked to Heart Defects in Babies, Cancer and Parkinson's. Republicans in Congress Want to Reverse a Ban on It," *ProPublica*, March 26, 2025; "Risk Management for Trichloroethylene (TCE)," Environmental Protection Agency, accessed March 24, 2025, epa.gov/assessing-and-managing-chemicals-under-tsca/risk-management-trichloroethylene-tce.

287. Sirena Rowland, "The Impact of the Military on Crime Rates in Towns Across the United States" (honors thesis, Pennsylvania State University, 2017), honors.libraries.psu.edu/files/final_submissions/3933.

288. Sexual Assault Prevention and Response Office, "Reports Archive," US Department of Defense, undated, https://www.sapr.mil/reports#tabs-0-57452200-1710862220-1.

289. *Annual Report on Sexual Assault in the Military: Fiscal Year 2022* (US Department of Defense, 2022), sapr.mil/sites/default/files/public/docs/reports/AR/FY22/DOD_Annual_Report_on_Sexual_Assault_in_the_Military_FY2022.pdf; *Annual Report on Sexual Assault in the Military: Fiscal Year 2021* (US Department of Defense, 2021), sapr.mil/sites/default/files/public/docs/reports/FY21_Annual_Report.pdf.

290. Vine, "United States"; other Okinawa studies: Takeshi Tokashiki, "Research on the Effect of the Air Craft Noise Pollution on the Noise Environment in the School Education of Okinawa Due to the US Military Bases," *INTER-NOISE and NOISE-CON Congress and Conference Proceedings* 253 (2016): 5918–26, https://www.ingentaconnect.com/content/ince/incep/2016/00000253/00000002/art00010 ; Takako Hikotani et al., "Revisiting Negative Externalities of US Military Bases: The Case of Okinawa," *International Relations of the Asia-Pacific* 23, no. 2 (2023): 325–49.

291. Graeber, *Bullshit Jobs*, 285.

292. Ibid (emphasis in original).

293. Milena Büchs, "Sustainable Welfare: How Do Universal Basic Income and Universal Basic Services Compare?" *Ecological Economics* 189 (2021): 107152, https://www.sciencedirect.com/science/article/pii/S092180092100210X.

CHAPTER VI: AN ANTHROPAUSE OF OUR OWN

294. Giorgos Kallis, *Degrowth* (Agenda, 2018), 124.
295. Parrique credits an article of mine for this statement, which I tend to use a lot; e.g., Stan Cox, "That Green Growth at the Heart of the Green New Deal? It's Malignant," *Resilience*, January 15, 2019, resilience.org/stories/2019-01-15/that-green-growth-at-the-heart-of-the-green-new-deal-its-malignant/.
296. Céline Keller, *Who Is Afraid of Degrowth?* (Flyeralarm, 2020), 152–53.
297. Jason Hickel et al., "Degrowth Can Work—Here's How Science Can Help," *Nature* 612, no. 7940 (2022): 400–403.
298. For decades, the environmental-justice movement has fought to end and reverse business's longstanding practice of siting dirty, dangerous industrial facilities, waste landfills, and other hazards near predominantly Black, Latino, and Indigenous residential areas. They have achieved some successes, but more often, they run into brick walls of official nonchalance or hostility. Title VI of the 1964 Civil Rights Act prohibits such racist, discriminatory practices, but EPA has been a highly reluctant enforcer. And in defiance of Title VI, red-state governments have ensured that companies can continue to target the communities. Lylla Younes, "A New Salvo in the Fight to Protect the 'Holy Grail' of Environmental Justice," *Grist*, September 20, 2024, grist.org/justice/fight-to-protect-title-vi-epa-holy-grail-environmental-justice. However, once in power, the Trump administration purged environmental justice from federal policy almost completely.
299. Kohei Saito, *Slow Down: The Degrowth Manifesto* (Astra House, 2024), 86, 163.
300. Ibid, 226.
301. Viviana Asara et al., "Degrowth, Democracy and Autonomy," *Environmental Values* 22, no. 2 (2013): 217–39.
302. Andrew Ahern, "Red and Green Make . . . Degrowth: On Kohei Saito's 'Marx in the Anthropocene'," *Los Angeles Review of Books*, July 23, 2023, lareviewofbooks.org/article/red-and-green-make-degrowth-on-kohei-saitos-marx-in-the-anthropocene/.
303. Kallis, *Degrowth*, 125.
304. Marty Logan, "World Social Forum Insists: Another World is Possible!" *Inter Press Service*, February 15, 2024, ipsnews.net/2024/02/world-social-forum-insists-another-world-possible.
305. Kallis, *Degrowth*, 162.
306. Elizabeth Derryberry et al., "Singing in a Silent Spring: Birds Respond to a Half-Century Soundscape Reversion During the COVID-19 Shutdown," *Science* 370 , no. 6516 (2020): 575–79.

AFTERWORD

307. US Energy Information Administration, "United States Produces More Crude Oil Than Any Country, Ever," March 11, 2024, eia.gov/todayinenergy/detail.php?id=61545.
308. Middle East Eye interview of Yanis Varoufakis, "The West Is Collapsing Under Its Own Lies," posted May 15, 2025, YouTube, 52 min., 21 sec., youtube.com/watch?v=vHtrLFwwVaI.

Index

STAN COX served as a research geneticist in the US Department of Agriculture in the 1980s and 1990s and as a senior scientist at The Land Institute from 2000 to 2024. Cox is the author of seven books, including *The Path to a Livable Future: A New Politics to Fight Climate Change, Racism, and the Next Pandemic*; *The Green New Deal and Beyond: Ending the Climate Emergency While We Still Can*; *Losing Our Cool: Uncomfortable Truths About Our Air-Conditioned World*; and *How the World Breaks: Life in Catastrophe's Path, from the Caribbean to Siberia* (with Paul Cox). His writing about the economic and political roots and consequences of the global ecological crisis have been published by *The New York Times*, *The Washington Post*, *Los Angeles Times*, *The Nation*, *The New Republic*, *Al Jazeera*, *Yes!*, *The Progressive*, and local publications spanning forty-three US states. In 2012, *The Atlantic* named Cox their "Readers' Choice Brave Thinker" for his critique of air conditioning. He lives in Salina, Kansas.